RECENT ADVANCES IN TIME SERIES FORECASTING

Manufacturing and Production Engineering Series

Series Editors: Hamid R. Parsaei, Texas A&M University at Qatar, & Waldemar Karwowski, University of Central Florida

This series will provide an outlet for the state-of-the-art topics in manufacturing and production engineering disciplines. This new series will also provide a scientific and practical basis for researchers, practitioners, and students involved in areas within manufacturing and production engineering. Issues envisioned to be addressed in this new series would include, but not limited to, the following: Additive Manufacturing, 3D Visualization, Mass Customization, Material Processes, Cybersecurity, Data Science, Automation and Robotics, Underwater Autonomous Vehicles, Unmanned Autonomous Vehicles, Robotics and Automation, Six Sigma and Total Quality Management, Manufacturing Cost Estimation and Cost Management, Industrial Safety, Programmable Logic Controllers, to name just a few.

Decision Making in Risk Management
Quantifying Intangible Risk Factors in Projects
Christopher Cox

Reconfigurable Manufacturing Enterprises for Industry 4.0
Ibrahim H. Garbie and Hamid Parsaei

Recent Advances in Time Series Forecasting
Dinesh C. S. Bisht and Mangey Ram

Cost Analysis for Engineers and Scientists
Fraiborz (Fred) Tayyari

For more information on this series, please visit: https://www.routledge.com/ Manufacturing-and-Production-Engineering/book-series/CRCMNPRDENG

RECENT ADVANCES IN TIME SERIES FORECASTING

Edited by
Dinesh C. S. Bisht and Mangey Ram

CRC Press
Taylor & Francis Group
Boca Raton London New York

CRC Press is an imprint of the
Taylor & Francis Group, an **informa** business

First edition published 2022
by CRC Press
6000 Broken Sound Parkway NW, Suite 300, Boca Raton, FL 33487-2742

and by CRC Press
2 Park Square, Milton Park, Abingdon, Oxon, OX14 4RN

© 2022 Taylor & Francis Group, LLC

CRC Press is an imprint of Taylor & Francis Group, LLC

Library of Congress Cataloging-in-Publication Data
A catalog record has been requested for this book

Names: Bisht, Dinesh C. S., editor. | Ram, Mangey, editor.
Title: Recent advances in time series forecasting / edited by Dinesh C.S.
Bisht and Mangey Ram.
Description: First edition. | Boca Raton : CRC Press, 2022. | Series:
Mathematical engineering, manufacturing, and management sciences |
Includes bibliographical references and index.
Identifiers: LCCN 2021021533 | ISBN 9780367607753 (hbk) | ISBN
9780367608699 (pbk) | ISBN 9781003102281 (ebk)
Subjects: LCSH: Time-series analysis. | Systems engineering--Mathematics. |
Communicable diseases--Transmission--Forecasting.
Classification: LCC QA280 .R44 2022 | DDC 519.5/5--dc23
LC record available at https://lccn.loc.gov/2021021533

ISBN: 978-0-367-60775-3 (hbk)
ISBN: 978-0-367-60869-9 (pbk)
ISBN: 978-1-003-10228-1 (ebk)

DOI: 10.1201/9781003102281

Typeset in Times
by MPS Limited, Dehradun

Contents

Preface

Future predictions have been a topic of interest since the existence of humans. These predictions can be seen in daily actions, for example, managing energy, air index predictions, stock prices, weather, total sales, etc. Precise estimates are crucial in monetary activities as forecasting errors in certain areas may lead to a big monetary loss. A time series is a chronological collection of data. The sequential analysis of data and information gathered from past to present is called time series analysis. Time series data are of high dimension, have a large size and are updated continuously. A time series depends on various factors like trend, seasonality, cycle and irregular data set. A time series is basically a series of data points well-organized in time. Time series forecasting is a significant area of machine learning. There are various prediction problems that are time dependent, and these problems can be handled through time series analysis. This book aims to cover the recent advancement in the field of time series analysis. It will cover theoretical as well as recent applications of the time series. The projected audience for this book is scientists, researchers and postgraduate students.

Editors

Dr. Dinesh C. S. Bisht received his Ph.D. with a major in mathematics and a minor in electronics and communication engineering from G. B. Pant University of Agriculture & Technology, Uttarakhand. Before joining the Jaypee Institute of Information Technology, he worked as an Assistant Professor at ITM University, Gurgaon, India. He has been a Faculty Member for approximately 11 years and has taught several core courses in applied mathematics and soft computing at undergraduate and master's levels. His major research interests include soft computing and nature-inspired optimization. He has published more than 35 research papers in national and international journals of repute. He is the Associate Editor for *International Journal of Mathematical, Engineering and Management Sciences, ESCI* and *SCOPUS* indexed journal. He is the editor of the book *Computational Intelligence: Theoretical Advances and Advanced Applications,* published by Walter de Gruyter GmbH & Co KG. He has also published seven book chapters in reputed book series. Dr. Bisht is a member of the International Association of Engineers in Hong Kong and the Soft Computing Research Society, India. He has been awarded for outstanding contribution in reviewing, from the editors of *Applied Soft Computing Journal,* Elsevier.

Dr. Mangey Ram received a Ph.D. in mathematics with a minor in computer science from G. B. Pant University of Agriculture and Technology, Pantnagar, India. He has been a Faculty Member for approximately 12 years and has taught several core courses in pure and applied mathematics at undergraduate, postgraduate and doctorate levels. He is currently the Research Professor at Graphic Era (deemed to be University), Dehradun, India. Before joining Graphic Era, he was a Deputy Manager (Probationary Officer) with Syndicate Bank for a short period. He is Editor-in-Chief of *International Journal of Mathematical, Engineering and Management Sciences* and *Journal of Reliability and Statistical Studies;* Editor-in-Chief of six book series with Elsevier, CRC Press, Taylor & Francis Group, Walter De Gruyter Publisher Germany and River Publisher; and the Guest Editor and Member of the editorial board of various journals. He has published more than 225 research publications (journal articles, books, book chapters, conference articles) in IEEE, Taylor & Francis, Springer, Elsevier, Emerald, World Scientific and many other national and international journals and conferences. Also, he has published more than 50 books (authored/edited) with international publishers like Elsevier, Springer Nature, CRC Press, Taylor & Francis Group, Walter De Gruyter Publisher Germany and River Publisher. His fields of research are reliability theory and applied mathematics. Dr. Ram is a Senior Member of IEEE and a Senior Life Member of Operational Research Society of India; Society for Reliability Engineering, Quality and Operations Management in India; and Indian Society of Industrial and Applied Mathematics. He has been a member of the organizing committee of a number of international and national conferences, seminars and workshops. He was conferred with the "Young Scientist Award" by the Uttarakhand State Council for Science and Technology, Dehradun, in 2009. He was awarded the "Best Faculty Award" in 2011; the "Research Excellence Award" in 2015; and the "Outstanding Researcher Award" in 2018 for his significant contribution in academics and research at Graphic Era (Deemed to be University), Dehradun, India.

1 Time Series Econometrics: Some Initial Understanding

A. Mishra
CHRIST (Deemed to be a University)

1.1 INTRODUCTION

In our daily lives, when we talk about a topic that is related to economics and finance, generally we relate it to time. Most of the macroeconomic variables, which have enormous significance in our lives, are measured in the time dimension. We are very familiar with these variables, such as GDP of a nation, inflation rate and unemployment rate. Likewise, in finance many things are calibrated on the time dimension, such as stock prices and return on investment. Measuring variables with regular frequency gives us a significant platform for useful analysis. Time series econometric analysis has massive implications for making inferences in dynamic circumstances; hence, its uses are surging day by day. These days, the innovations of machine learning and data analysis are expanding the need for time series econometric analysis beyond the dimension of economics and finance. For example, time series analysis data of a cardiac patient can be used to forecast the exact time of a possible cardiac attack in the future through the application of various sophisticated time series econometric modeling.

1.1.1 LEARNING OBJECTIVES

After completing this chapter, students should be able to:

- Understand the meaning of the time series.
- Understand the random walk phenomenon.
- Recognize the stationarity and unit root process.
- Understand the concept of spurious regression.
- Identify the relevancy of a stationary time series.

1.2 TIME SERIES, WHAT IS IT?

When we say "time series", one thing is obvious—it has two dimensions. The first is related to time, and the second is that it is a series, which is a collection of variables. Hence, a time series can be defined as accumulating random variables according to time or in chronological order.

DOI: 10.1201/9781003102281-1

1

In the time series, the quantitative characteristic is collected or arranged with an equal time interval. For example, suppose we are arranging the GDP growth of a nation in years. Then, we cannot change the time interval to months. We have to follow the same sequence of time for each variable. If we follow the same time sequence, the series will be considered a time series of data. One important aspect that we have to keep in mind is that the time interval could be any period; it may be yearly, quarterly, monthly, weekly, daily or even hourly. Accordingly, we should classify the data as a yearly time series, monthly time series, weekly time series or daily time series.

Figure 1.1 shows a graph of daily trading time series data for the Indian stock exchange from 2010 onwards.

1.2.1 FOUR COMPONENTS OF A TIME SERIES

Usually, any time series includes the four components of trend, cyclical, seasonal and irregular, which are discussed below. These components have their relevancy in the forecasting process. When analyzing the time series data from a statistical point of view, we always pay attention to these four components, and our primary intention is to decompose these components so that we can forecast. There are two ways to deal with these components, the addition method and the multiplication method. We now discuss the components of a time series in detail.

1.. Trend (T)
2.. Cyclical (C)
3.. Seasonal (S)
4.. Irregular (I)

1.2.2 TREND COMPONENT

The trend is the long-run array of a time series. A trend can be upward or downward, conditional on whether the time series exhibits a surging long-term outline or

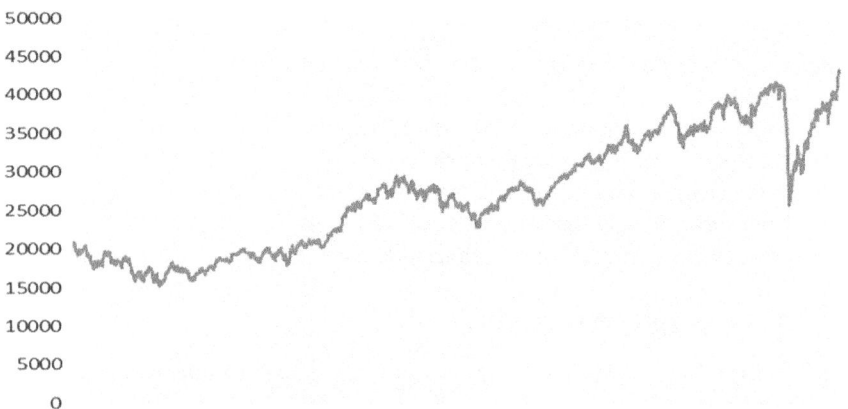

FIGURE 1.1 Indian stock exchange, Sensex daily trading data from 2010 onward.

a diminishing long-term array. On the other hand, if a time series does not demonstrate an upward or downward outline, it is considered stationary in the mean.

1.2.3   CYCLICAL COMPONENT

Sometimes there is a continuous fluctuation in the trend line. Sometimes it moves up, and sometimes it falls. Such continuous movement is considered a cyclical pattern. The length of a cyclical pattern is determined by the type of industry and business that has been selected for the analysis.

1.2.4   SEASONAL COMPONENT

Seasonality happens when the time series shows systematic variations throughout an identical time sequence. The time sequence could be monthly or quarterly each year. For instance, there is a market sale surge throughout the month of Diwali and the festive session in India and during Christmas in Christian parts of the world.

1.2.5   IRREGULAR COMPONENT

The irregular element of a time series will always be random and impulsive in nature. Every time series has some distinctive constituent that transforms it into an unsystematic variable. In forecasting, the main motto is to "calibrate" each and every component of the time series in a precise manner except for the irregular component. The invention of machine learning and the econometrics model made it possible to measure the unsystematic variable or irregular component.

$$y_t \quad where \ t = 1, \ 2, \ 3, \ \dots\dots\dots\dots\, n$$

Generally, when we represent the time series data, variables are attached with the subscript **t,** where **t** could be any natural number. The letter **t** represents the value of that variable at the particular time period. Suppose we say y_5 means that we are talking about the value of Y at the fifth time period. The random variables can be represented in the form **y(t)** instead of y_t. Now, the question emerges: What is the difference between these two representations? The notation **y(t)** indicates that the concerned random variable is continuous whereas the notation y_t indicates a discrete random variable.

1.2.6   TIME SERIES IN ECONOMETRIC ANALYSIS

Time series analysis in econometrics involves many technical words we would like to define. We start with the term *stochastic processes.* It is the mechanism of collecting random variables according to time. The term *stochastic processes* is frequently applied in time series econometrics analysis. These processes include various forms, such as stationary stochastic processes, weakly stationary stochastic processes and non-stationary stochastic processes, which we will discuss next.

1.3   STATIONARY STOCHASTIC PROCESSES

When we discuss time series econometrics analysis, our journey begins with the concept of stationary stochastic processes. We have already discussed the term *stochastic processes*; now, there is one additional term before that—*stationary*.

> *"a stochastic process is said to be stationary if its mean and variance are constant over time and the value of the covariance between the two time periods depends only on the distance or gap or lag between the two time periods and not the actual time at which the covariance is computed"Gujarati and Porter (2003).*

Through the above definition, we learn that when we talk about stationary stochastic processes, we need to look at the mean, the variance and the covariance of the random variables. For the stochastic processes to be stationary, the mean and variance should be constant over time while the covariance should depend on the gap between the two time periods where the calculation is made. In econometrics, such a stochastic process is also known as a weakly stationary stochastic process.

Figure 1.2 reveals the graph of non-stationary time-series data. The stochastic process is not following the stationarity property as the data are not moving around the mean; therefore, we can conclude that the mean is time variant and not a stationary stochastic process. On the other hand, Figure 1.3 shows the graph of a stationary stochastic process, as the data set is ultimately moving around its mean. The graphical representation of data gives an initial clue about the stationary stochastic process.

The above term can be represented with a simple notation. Suppose we are talking about the random variable Z_t as a stochastic process with the following attributes.

$$\text{Mean } Expectation \ (Z_t) = \mu$$

FIGURE 1.2   Non-stationary time series data.

FIGURE 1.3   Stationary time-series data.

$$\textbf{Variance } \textit{var } (Z_t) = (Z_t - \textit{Expectation } (Z_t))^2 = \sigma^2$$

$$\textbf{Covariance } \vartheta_l = E[(Z_t - E(Z_t))(Z_{t+l} - E(Z_t))]$$

From the above notation, it can be seen that ϑ_l, the covariance (or auto covariance) at lag l time period, is the covariance between the values of Z_t and Z_{t+l}, that is, between two Z *values* l periods apart. If $l = 0$, we obtain ϑ_l, which clearly demonstrates that variance of $Z = \sigma^2$. The above equation also reveals that the mean and variance of random variable Z is μ *and* σ^2, which is time invariant and remains constant; this is exclusive to the stationary stochastic process. Now, suppose we shift the origin of random variable Z from the time period *t* to any time *u*. Then, a question arises: What will happen to the stationarity property of the variable Z? In this case, also, our stochastic process will continue to follow the property of stationarity. It will be a stationary stochastic process. It further documents that no matter how much the time changes, this type of stochastic process will deviate around its mean and maintain the stationarity property.

There is an important question that may arise in the mind of the reader: Why does a stationary time series deviate around its mean? The answer lies in the stationarity property of time series, where the variance is also time invariant. The constant variance does not allow the mean to drift away too far from its original position. Is this the case for each stochastic process; meaning, does every stochastic process follow the stationary attributes? The answer is no. In reality, all time series are not stationary, and those that are not stationary are recognized as non-stationary stochastic processes. In econometrics, there are various methods and tools available to comment about the stationarity attributes of any stochastic process. If any time series is non-stationary, it means that it will have a time-varying mean or a time-varying variance or both, and that is not desirable for analysis. We will shortly understand why it is necessary to have a stationary time series at the end of the current chapter.

1.4 RANDOM WALK PHENOMENON IN TIME SERIES

The term *random walk* is a vaguely familiar concept; we hear this term in finance. The random walk is usually compared with a drunk person. If we look at the footsteps of a drunk person, we may realize that his footsteps are not consistent, and when he leaves the wine shop, he moves away and away in an inconsistent manner from the shop from where he has drunk. Why are we talking about such things here? What is the relevancy of it? How is the random walk concept related to the stock market or financial economics? Let us discuss it in a more comprehensive way. Consider the stock market; the price of a share in the financial market shows random walk behavior. In the stock market generally, today's share price is equivalent to last trading day's share price plus a random shock. If we analyse the random walk phenomenon, we will find that it generates a non-stationary stochastic process. The random walk concept is not only related to stock markets, but it can also be experienced with the exchange market, money market and so forth. Let us now discuss why the random walk model creates a non-stationary series. First, we talk about random walk without drift, then random walk with drift.

Let us create a model:

$$Z_t = Z_{t-1} + e_t \tag{1.1}$$

Suppose e_t is a residual term (error) which has zero mean and constant variance (σ^2). In this case, the given model will be considered a random walk without drift. The equation (1.1) demonstrates that there is not an intercept term, which indicates that the current value of Z is determined by its own, one lag period value and some random shocks, which is **iid**[1]$(0, \sigma^2)$; hence, the above model is considered a random walk without drift.

Now, equation (1.1) can be written as

$$Z_1 = Z_0 + e_1$$
$$Z_2 = Z_1 + e_2, \quad Z_2 = Z_0 + e_1 + e_2$$
$$Z_3 = Z_2 + e_2, \quad Z_3 = Z_2 + e_3, \quad Z_3 = Z_0 + e_1 + e_2 + e_3$$

If we follow the same process, we will get

$$Z_t = Z_0 + \sum e_t \tag{1.2}$$

Finally, if we have the expectation of Z_t, then it would be

$$E(Z_t) = E\left(Z_0 + \sum e_t\right)$$

$$E(Z_t) = E(Z_0) + E\left(\sum e_t\right)$$

$E(Z_t) = Z_0$, (Because the expectation of the sum of the error term is zero.)
    Similarly, if we calculate the variance of Z_t, it will be

$$\text{var } (Z_t) = t\sigma^2$$

This is because the variance of the error term is independently identically distributed (iid); therefore, each error term varies corresponding to its time period, and the variance terms are time variant. In the random walk model without drift, the mean is constant or time invariant, but the variance term is not constant and time variant. Therefore, it is considered a non-stationary stochastic process.

1.4.2 RANDOM WALK WITH DRIFT

We have learned that the random walk model without drift is a non-stationary stochastic process. Let us move to extend our understanding to the random walk model with drift. We will modify the equation (1.1) by adding an additional term intercept or drift:

$$Z_t = \beta + Z_{t-1} + e_t \tag{1.3}$$

The equation (1.3) includes the intercept or drift term (β) in the model, and the rest of the things are the same as in the equation (1.2). Because we have added the intercept term, the above model is called the random walk model with drift. Now, the question arises: Does the random walk model follow all the properties of stationarity or is it the same as we have seen in the case of the random walk model without drift?
    If we follow the same process we followed in the case of the random walk without drift, we will find that the random walk with drift model is also a non-stationary stochastic process. The brief explanation has been given below.

$$E(Z_t) = Z_0 + t\beta$$

Similarly, if we calculate the variance of Z_t, it will be

$$\text{var } (Z_t) = t\sigma^2$$

We clearly see that in the case of the random walk model with drift, mean as well as variance are time variant, which violates the property of stationarity. Therefore, the above-mentioned random walk model with drift and without drift both are non-stationary stochastic processes.

1.4.3   Unit Root Stochastic Process

The unit root process has become one of the central concepts in time series econometric analysis. It is an effective mechanism of explanation for the stationarity of any time series. Let us write a model:

$$Z_t = \rho Z_{t-1} + e_t \tag{1.4}$$

In the above equation, it can be seen that if the value of $\rho = 1$, then the equation (1.4), will automatically become the random walk model without drift, and we have already discussed that the random walk model is a non-stationary stochastic process. On the other hand, if $|\rho| < 1$, let us assume that if the value of ρ falls less than 1, it can be revealed that Z_t is a stationary stochastic process in a sense based on the property that we have seen it. On the other hand, if our assumption of the equation (1.4) is that the preliminary value of Z_0 is nil, $|\rho| < 1$ and e_t is white noise and normally distributed, $N(0, \sigma^2)$; in this case, the expected value of Z is $E(Z_t) = 0$ and $var(Z_t) = 1/(1 - \rho^2)$. Subsequently, it can be surmised that mean and variance seem to be time invariant; in other words mean and variance are constant. Hence, by explanation of the weak stationarity process, Z_t is a stationary stochastic process. In other circumstances, we have already seen before, if $\rho = 1$, Z_t is a random walk or non-stationary stochastic process. Hence, in general, it is essential to find out if a stochastic process possesses a unit root or not. We will discuss the statistical test which deals with the unit root process of time series.

1.5   SPURIOUS REGRESSION IN TIME SERIES ANALYSIS

Many times in time series analysis we misunderstand the outcome of regression that has been generated by running a regression on two or more than two time series. Let us take an example to understand the concept of spurious regression. Suppose we have two different stochastic processes, **A and B,** with 500 observations each. Let us also assume that these time series are non-stationary, but we are not aware of this fact, and we regress A on B.

If we analyze the outcome of regression, we see that the regression coefficient is highly significant, which indicates that there is a strong relationship among the variables A and B. On the other hand, the *coefficient of determination* R^2 value is small; it is considerably dissimilar from zero. On the basis of these results, we can easily conclude that there is a significant statistical relationship between A and B; however, there should not be any relationship between these two variables. In reality, these two time series are non-stationary; therefore, there cannot be a logical relationship, but our regression result reveals that there is a substantial arithmetical association between A and B. Where is the problem? The answer to this question is the concept of spurious regression. Actually, in this case, we are running a spurious regression. One thing which we ignored from the regression result is Durbin–Watson **d** statistics. If we look at the d statistics, we find that it is less than R^2. This is an indication of spurious or nonsense regression, first revealed by Yule (1926). In the context of spurious regression, Yule (1926) suggested a rule of thumb. If the value of R^2 is more than the Durbin–Watson **d** statistics, we may suspect spurious regression. If we regress one time series on the other, although there is not a theoretical relationship, the regression outcome may show a statistically significant relationship between the variables. We should be cautious about such outcomes and always keep this theory in mind.

Here is an example of spurious regression. Suppose we run the regression on the Egyptian infant mortality rate and money supply growth rate of India. The outcome of regression may show a significant relationship between Egyptian infant mortality rate and money supply growth rate of India, but is it logically true? How can Egyptian infant mortality rate and money supply growth rate of India be related? This melodramatic example shows that one should be cautious of conducting regression analyses based on time series that exhibit stochastic trends.

1.6 NEED FOR STATIONARY DATA

So far, we have discussed many concepts related to the time series process. Among all discussed topics, we emphasized most the stationary stochastic process. Now, the pertinent question emerges: Why is it necessary to have stationary time series data? The answer to this question lies within the property of the stationary stochastic process, where the mean and variance of the stochastic process are constant and do not vary with time, and the covariance is dependent on the gap between the two time intervals for which we are calculating the covariance. Let us look at the condition of a stationary process. It is clear that if a stochastic process is non-stationary, we can study its comportment simply for the time period under contemplation.

For the non-stationary stochastic process, it is not conceivable to specify it to other time periods. Hence, forecasting and making inferences with a non-stationary stochastic process may be of diminutive importance. A constant means and variance allow us to generalize the analysis to the real data set. Hence, we always look at the stationarity property of stochastic process before making any inferences with it.

NOTE

1 Independently identically distributed.

REFERENCES

Gujarati, D. N. & Porter, D.C. (2003). *Basic econometrics*. McGraw Hill Book Co.
Yule, G. U. (1926). Why do we sometimes get nonsense correlations between time series? A study in sampling and the nature of time series. *Journal of the Royal Statistical Society*, 89(1), 1–64.

2 Time Series Analysis for Modeling the Transmission of Dengue Disease

A.M.C.H. Attanayake[1] and S.S.N. Perera[2]
[1]Department of Statistics & Computer Science, University of Kelanya, Sri Lanka
[2]Research & Development Centre for Mathematical Modelling, Department of Mathematics, University of Colombo, Sri Lanka

2.1 INTRODUCTION

Dengue is a viral disease spread through humans by the bites of infected mosquitoes. The causing virus has been identified as DENV (dengue virus), and the four serotypes of the virus complicate the transmission possibilities of the disease. It is one of the fastest spreading diseases, and no exact treatment has been identified as a cure for the disease.

Approximately half of the population in the world is now facing the threat of dengue infection. More than 100 countries in the world are continuously infected with the disease. Globally, a growing number of mortalities and morbidities were reported annually due to severe dengue. From the beginning of 2020 to May 2020, over 1.4 million dengue cases have been reported in the world while reporting more than 600 deaths. As of May 15, 2020, the highest number of dengue cases reported occurred in Brazil, which was around 0.8 million, whereas the highest number of deaths was in the Philippines, approximately 300 people. Sri Lanka and Indonesia are on the list of affected countries with dengue disease. The map of Sri Lanka is shown in Figure 2.1, and the position of Colombo is indicated on the map. The latitude and longitude of Colombo are 6.927079 and 79.861244, respectively. Colombo is the commercial capital of Sri Lanka, and the majority of the public transports in the country are connected to the city of Colombo.

Dengue was found in Sri Lanka in 1960 (Sirisena & Noordeen, 2014). Afterwards, the burden of the disease amplified exponentially during the last decade. Generally, the highest number of dengue cases is reported in the Western Province of Sri Lanka, which is the most populated province among the nine provinces of the country. Within the Western province, the majority of infections

DOI: 10.1201/9781003102281-2

FIGURE 2.1   The position of Colombo in Sri Lanka.

were recorded in the Colombo district: 27,886 dengue cases reported in Sri Lanka from January to September 2020 (Epidemiology Unit, 2020). Sri Lanka had an explosive outbreak of dengue disease in the year 2017, reporting 186,101 total identified cases. During the year, island-wide precautions as well as preventive measures were implemented to reduce the impact of the disease. The distribution of dengue cases from 2010 to 2019 in Sri Lanka is depicted in Figure 2.2. Sri Lanka has experienced a remarkable growth in reported dengue cases over recent years.

Dengue was found in Indonesia in 1968 (Setiati et al., 2006), and presently the epidemic is apparent in all provinces of the country. Bali, Jakarta and East Java are provinces of Indonesia which have frequently reported dengue cases. The number of dengue cases recorded in Indonesia from January 1, 2019 to July 1, 2019 were 105,122 cases and 727 deaths. According to dengue case figures in 2015, Jakarta was the sixth province in the country (Ekasari et al., 2018). The distribution of dengue cases from 2008 to 2016 in Jakarta is depicted in Figure 2.3. The number of recordings were higher in 2016 than previous years. Jakarta is the largest city and the capital of Indonesia (Figure 2.4). The latitude and longitude of Jakarta are 6.2088 and 106.8456, respectively. Hence, the transmission of dengue disease is highly vulnerable in Jakarta.

Since dengue is a life-threatening and widely communicable disease, the introduction and implementation of controlling actions play a vital role in reducing the burden of the disease. Predictions generated by the models may be helpful in

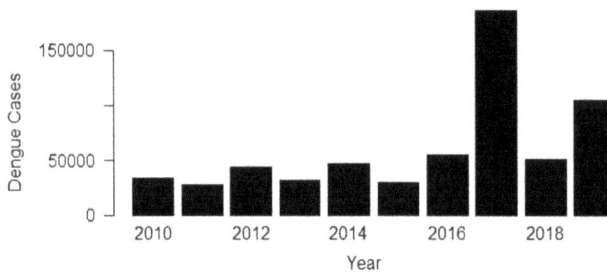

FIGURE 2.2   The distribution of the total number of dengue cases in Sri Lanka.

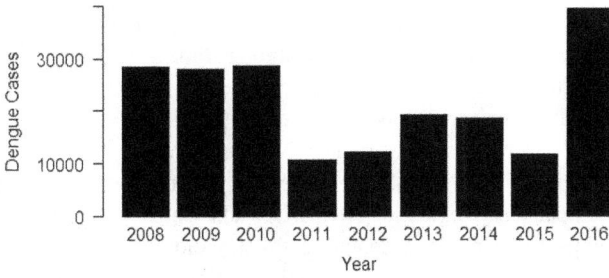

FIGURE 2.3 The distribution of the total number of dengue cases in Jakarta, Indonesia.

Jakarta

FIGURE 2.4 The position of Jakarta in the map of Indonesia.

executing controlling actions at the most appropriate period as well as to understand the underlying processes. Therefore, modeling and predicting are holding a prime position in the field of dengue management.

Numerous approaches for dengue modeling and predicting can be found in the area of statistics. In particular, time series analysis is widely used and one of the prominent approaches available in the field. Time series analysis demonstrates the methods available for investigations on a sequence of observations or results gathered in equal time gaps. Examples for time series include monthly rainfall in a certain city, annual population size, daily closing price of a stock index, etc. The time series analysis can be divided into frequency and time domain methods; in the time domain method, analysis of mathematical formulas and functions is based on time whereas the frequency domain relies on frequencies. Wavelet analysis is an example of a frequency domain analysis, and autocorrelation analysis is an example of the time domain category. Furthermore, time series analysis can be categorized as non-parametric and parametric methods. The former category does not assume any particular structure or pattern for the underlying process, whereas certain distributional patterns are assumed in the latter case. Univariate and multivariate analysis of time series are alternative areas in time series analysis. Univariate time series analysis describes the patterns or structures exhibited in one series, and multivariate time series analysis demonstrates more than a single time series at a particular time. The time series analysis has been extensively used in the early discovery of unexpected as well as explainable patterns of any infectious disease. The analysis highly relies on the assumption that the present behavior follows the patterns of the previous behaviors. The powerful techniques accessible through time series analysis can be applied to model the dengue incidence effectively and efficiently. The present chapter discusses some of the common univariate and multivariate techniques available in time series analysis such as Autoregressive Integrated Moving Average (ARIMA), exponential smoothing, decomposition, Alpha-Sutte modeling, Autoregressive Integrated Moving Average with Explanatory variables (ARIMAX) and exponential smoothing with explanatory variables using reported dengue cases in Colombo, Sri Lanka, and Jakarta, Indonesia, as test cases. Additionally, several combining forecasting approaches and improvement of forecasting results with ARIMA and exponential smoothing using combining approaches are illustrated.

Usage of ARIMA models in modeling and predicting dengue case counts can be found in the literature (Dom, 2013; Earnest et al., 2012; Eastin et al., 2014; Promprou et al., 2006). Many researchers utilized data from several years, and in some cases seasonal fluctuations were exhibited in the data series, emphasizing the appropriateness of Seasonal ARIMA (SARIMA) models. W. Anggraeni and L. Aristiani in 2016 demonstrated that an ARIMAX model using the Google Trends search index as an independent variable for dengue fever cases outperformed the ARIMA model on dengue fever cases in Indonesia (Anggraeni & Aristiani, 2016). Attanayake et al. (2020) published an article on predicting dengue cases in Colombo, Sri Lanka, using exponential smoothing. They concluded that exponential smoothing was an appropriate method for short-term prediction of dengue cases in the city they considered. Furthermore, Suneetha et al. (2018) and Ho and Ting (2015) explained the applications of exponential smoothing in modeling dengue incidence. Polwiang applied time series decomposition with the aim of describing the seasonal, trend and random components existing in the dengue series in Bangkok in 2020. Application of the same method for two Brazilian cities—Recife and Goiania—can be found in Cortesa et al. (2018).

The Alpha-Sutte indicator method applies to forecast data in various fields, including time series data. The method originated from the Sutte indicator (Ahmar et al., 2017). To the best of our knowledge, no applications of the Alpha-Sutte indicator method were found in the literature related to dengue modeling. The capability of the method in modeling dengue cases in Colombo, Sri Lanka, and Jakarta, Indonesia, is justified in this chapter. Furthermore, applications of exponential smoothing with covariates in the context of dengue disease are limited in the literature. To advance the forecasts generated by the ARIMA and exponential smoothing models as well as to remove unsatisfactory performances, multiple combined forecast approaches were tested by Attanayake et al. (2019); they concluded that the OLS regression method was the most appropriate combining forecast method to predict dengue cases recorded in Colombo, Sri Lanka.

The chapter is organized as follows. The first section gives a general introduction to the dengue disease, giving special attention to Colombo and Jakarta. It outlines methods in the time series analysis and provides a brief literature review. The next section explains seven techniques for time series analysis and their respective applications. Data from Jakarta and Colombo are used to illustrate the applications of time series methods. The conclusion and discussion segment summarizes the advantages and drawbacks of time series approaches by emphasizing the importance and effectiveness of combining modeling approaches as a current trend of modeling over the classical approaches mentioned.

2.2 THEORY AND APPLICATIONS

2.2.1 AUTOREGRESSIVE INTEGRATED MOVING AVERAGE (ARIMA) METHOD

ARIMA is one of the popular and widely adopted time series analysis methods. ARIMA derived from AR (autoregressive) and MA (moving average) methods. AR modeling is based on the fact that the recent value of the series is a function of past values. The MA is a function of past values of residuals. The ARIMA model is usually denoted as ARIMA (p, d, q), where p is the number of autoregressive components, q is the number of moving average components and d is the number of times that the original series is differenced. If no differencing is required to achieve a stationary process, then the ARMA model is appropriate. The model equation of ARMA (p, q) is given as:

$$Y_t - \varphi_1 Y_{t-1} - \cdots - \varphi_p Y_{p-1} = \varepsilon_t + \theta_1 \varepsilon_{t-1} + \cdots + \theta_q \varepsilon_{q-1} \qquad (2.1)$$

where ε_t is the white noise and Y_t is the series. Using the lag operator L, the autoregressive and moving average polynomial is represented as $\varphi(L)$ and $\theta(L)$.

The model equation of ARIMA (p, d, q) is given as:

$$\varphi(L)(1 - L)^d Y_t = \theta(L)\varepsilon_t \qquad (2.2)$$

where d is the dth difference operator.

After the time series plot is examined, stationarity of the series can be tested by applying the Augmented Dickey Fuller test (ADF), Kwiatkowski–Phillips–Schmidt–Shin (KPSS) test or Phillips-Perron (PP) test. These tests are capable of detecting whether a series is a stationary series or not. The null hypothesis for the ADF and PP tests is "series is non-stationary" whereas for the KPSS test, it is "stationary". If the original series is not stationary, then it can be transformed into a stationary one by differencing the data or using some transformation. The autocorrelation function (ACF) and partial auto-correlation function (PACF) of a stationary series should be inspected when determining the order of processes in the AR and/or MA model or mixed ARIMA model. The most appropriate model is then selected based on accuracy measures such as Akaike Information Criteria (AIC), Corrected Akaike Information Criteria (AICc) and Bayesian Information Criteria (BIC) measures. In model diagnostic checking, residuals of the selected model followed a white noise which is drawn from a constant mean and variance. If the assumptions are not held, then another model needs to be investigated; otherwise, the model can be used to make predictions after validation (Box & Jenkins, 1970). The mean absolute percentage error (MAPE) is less than 10%, revealing that the model is appropriate in forecasting (Lewis, 1982). MAPE criteria for model evaluation is summarized in Table 2.1.

The Seasonal Autoregressive Integrated Moving Average (SARIMA) method is utilized in the cooperating seasonal component in ARIMA modeling. The SARIMA model is denoted as SARIMA $(p, d, q) (P, D, Q)_m$, where m is the number of periods

TABLE 2.1
MAPE Criteria for Model Evaluation
(Lewis, 1982)

MAPE (%)	Forecasting Power
<10	High Accurate Forecasting
10–20	Good Forecasting
20–50	Reasonable Forecasting
>50	Inaccurate Forecasting

in each season. P, D, and Q are autoregressive, differencing and moving average components of the seasonal part. The seasonal part of an autoregressive (AR) and/or moving average (MA) model is identified at the seasonal lags of the PACF and ACF plots. The modeling procedure is almost the same as for the ARIMA procedure.

Note that "tseries" and "forecast" packages of R software can be used to fit ARIMA or SARIMA models.

2.2.1.1   Application of ARIMA/SARIMA Modeling

We used monthly reported dengue cases in Jakarta, Indonesia, from January 2010 to December 2015 for the analysis. The time series plot of dengue cases is shown in Figure 2.5. The reported cases vary between 307 and 4,121, with the mean number of cases as 1,447. The peak of the distribution is apparent in March, 2010. Seasonal variations exhibit in the series.

ACF and PACF plots are depicted in Figure 2.6 and Figure 2.7. In the ACF plot, the first three lags were significant, and some seasonal lags were significant around 18, whereas in the PACF plot, the first lag was significant with significant seasonal lag at 13. Since there are apparent seasonal lags, the SARIMA models may be more appropriate. Both PP and KPSS tests confirmed the stationarity of the original series at the 5% level of significance (p values were 0.037 and 0.1, respectively). We tested different candidate models, and summary measures are given in Table 2.2.

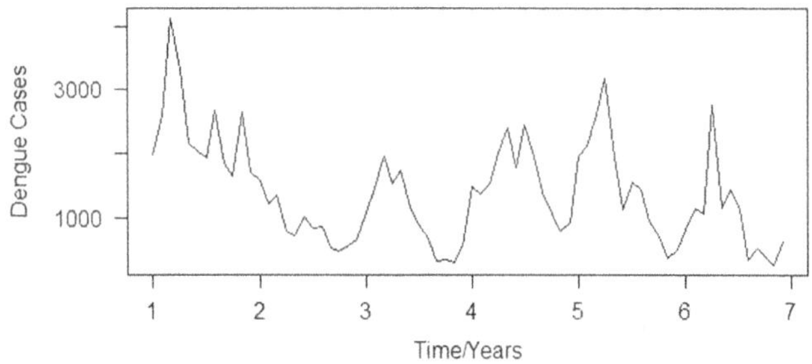

FIGURE 2.5   The time series plot of dengue cases in Jakarta, Indonesia.

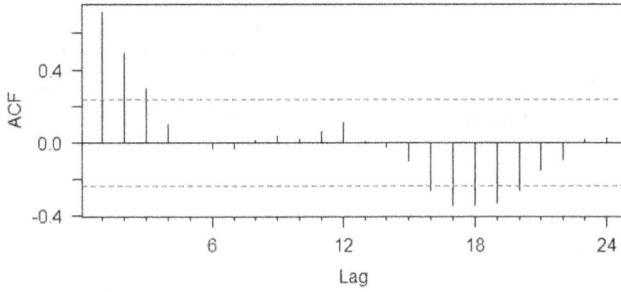

FIGURE 2.6 ACF plot of original Series—Jakarta, Indonesia.

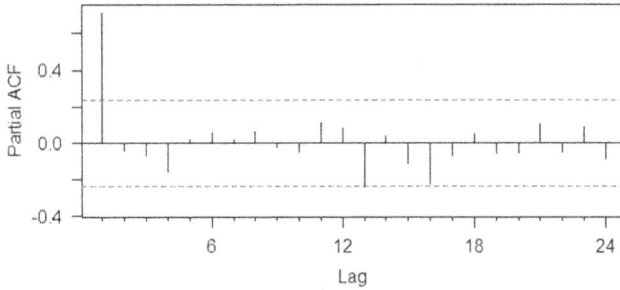

FIGURE 2.7 PACF plot of original series—Jakarta, Indonesia.

TABLE 2.2
Candidate ARIMA/SARIMA Models

Model	AIC	AICc	BIC
ARIMA $(2,0,2)(1,0,1)_{12}$	1074.07	1076.47	1091.95
ARIMA $(1,0,0)(1,0,0)_{12}$	1066.41	1067.03	1075.34
ARIMA $(0,0,1)(0,0,1)_{12}$	1087.53	1088.15	1096.46
ARIMA $(1,0,0)$	1073.32	1073.69	1080.02
ARIMA $(1,0,0)(1,0,1)_{12}$	1068.27	1069.22	1079.45
ARIMA $(1,0,0)(0,0,1)_{12}$	1068.02	1068.65	1076.96
ARIMA $(0,0,0)(1,0,0)_{12}$	1122.36	1122.73	1129.07
ARIMA $(2,0,0)(1,0,0)_{12}$	1068.39	1069.35	1079.57
ARIMA $(1,0,1)(1,0,0)_{12}$	1068.39	1069.35	1079.58
ARIMA $(0,0,1)(1,0,0)_{12}$	1088.17	1088.79	1097.12
ARIMA $(2,0,1)(1,0,0)_{12}$	1070.87	1071.68	1083.73

The most appropriate SARIMA model which has minimum summary measures was seasonal ARIMA $(1,0,0)$ $(1,0,0)_{12}$. Next, we tested the model for assumptions. We used the Jarque Bera Test to validate the normality assumption of residuals whereas we used the Box-Ljung test on residuals and on squared residuals to

validate the no autocorrelations and constant error variance assumptions, respectively. All the assumptions were satisfied by the selected model at 5% significance
levels by reporting p values as 0.062, 0.8796 and 0.4801 for the Jarque Bera test,
Box-Ljung test on residuals and Box-Ljung test on squared residuals, respectively.
A graphical summary of residual analysis is shown in Figure 2.8.

We validated the most appropriate model using the last three months' data, and
the corresponding MAPE was 52.6%. Therefore, according to Table 2.1, although
the selected model is the best ARIMA model, it is not adequate for forecasting
dengue cases in Jakarta.

2.2.2 AUTOREGRESSIVE INTEGRATED MOVING AVERAGE WITH EXPLANATORY VARIABLES (ARIMAX)

ARIMAX is an advanced approach for ARIMA modeling that supports investigating
the influence of independent variables (covariates) on the dependent variable.

The model equation of ARIMAX (p, d, q) is given as:

$$\varphi(L)(1 - L)^d Y_t = \theta(L)\varepsilon_t + \vartheta(L)X_t \tag{2.3}$$

where X_t denotes independent variables and $\vartheta(L)$ denote the coefficients.

If seasonal fluctuations exist, then the seasonal ARIMAX (SARIMAX) model
would be appropriate. The overall modeling procedure is similar to ARIMA
modeling except explanatory variables exist and need to find cross-correlations of
independent variables with a dependent variable. Then, you need to couple those
with the main building procedure; "xreg" parameter of the "arima" function in the
"forecast" package supports ARIMAX/SARIMAX models.

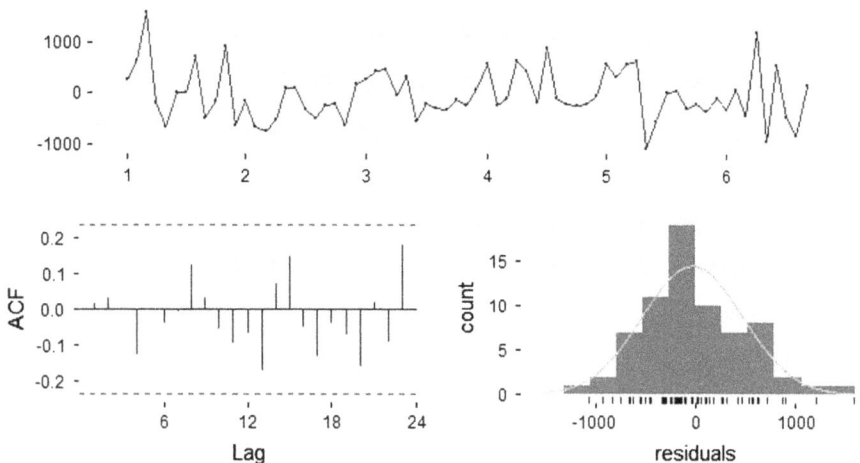

FIGURE 2.8   Residual analysis—ARIMA (1,0,0) (1,0,0)$_{12}$.

2.2.2.1   Application of ARIMAX/SARIMAX

We used monthly reported dengue cases in Jakarta, Indonesia, from January 2010 to December 2015 for the analysis. The last three months' data were used to validate the modeling procedure. We selected average monthly rainfall and average monthly humidity data from January 2010 to December 2015 as independent variables that influence the dengue distribution in Jakarta.

Cross-correlation results are summarized in Table 2.3. According to Table 2.3, two months leading rainfall and humidity were associated with monthly dengue cases in Jakarta.

In addition, Pearson correlations of covariates with associated lag periods and dengue cases were statistically not significant at a 5% level of significance (Table 2.4).

Therefore, we modeled monthly dengue cases with the covariates using an ARIMA with explanatory variables method. A summary of the fitted ARIMAX models is shown in Table 2.5.

The lowest AIC, AICc and BIC values are recorded for the ARIMAX model with rainfall as an independent variable. We verified properties of the residuals of

TABLE 2.3
Cross Correlation of Variables

Lag in Months	Cross Correlations	
	Rainfall and Dengue	Average Humidity and Dengue
10	−0.242684	−0.115583
9	−0.248579	−0.166221
8	−0.211648	−0.191189
7	−0.138680	−0.207165
6	0.000377	−0.079237
5	0.098028	0.048011
4	0.274262	0.270023
3	0.453892	0.516530
2	0.587331	0.603961
1	0.383866	0.594639
0	0.166335	0.400070

TABLE 2.4
Pearson Correlations

	Lag in Months	Pearson Correlation	P Value
Dengue and Rainfall	2	0.596	0.000
Dengue and Humidity	2	0.616	0.000

TABLE 2.5

Summary Measures of Fitted ARIMAX Models

Covariate/s in the Model	AIC	AICc	BIC
Rainfall	1040	1041	1051
Humidity	1051.9	1052.6	1060.8
Rainfall, Humidity	1041.9	1043.4	1055.3

this model by applying the Ljung-Box test on residuals as well as on squared residuals at the 5% level of significance (corresponding p values were 0.5559 and 0.741, respectively). According to the Jarque Bera test, residuals followed a normal distribution, and p value of the test recorded as 0.2144. Graphical representation of residuals is shown in Figure 2.9.

As the assumptions were held by the selected ARIMAX model, it is the most appropriate model. The model is seasonal ARIMAX $(1,0,0)$ $(1,0,0)_{12}$. The MAPE value of the validation set is 53.6%. Therefore, according to Table 2.1, although the selected model is the best seasonal ARIMAX model, it is not adequate in forecasting dengue cases in Jakarta.

2.2.3 EXPONENTIAL SMOOTHING

In this smoothing method, the latest observations are weighted more deeply than earlier observations (Attanayake et al., 2020). The "ets" function of the "forecast" package which is available in the R software package can be used to fit an exponential smoothing model.

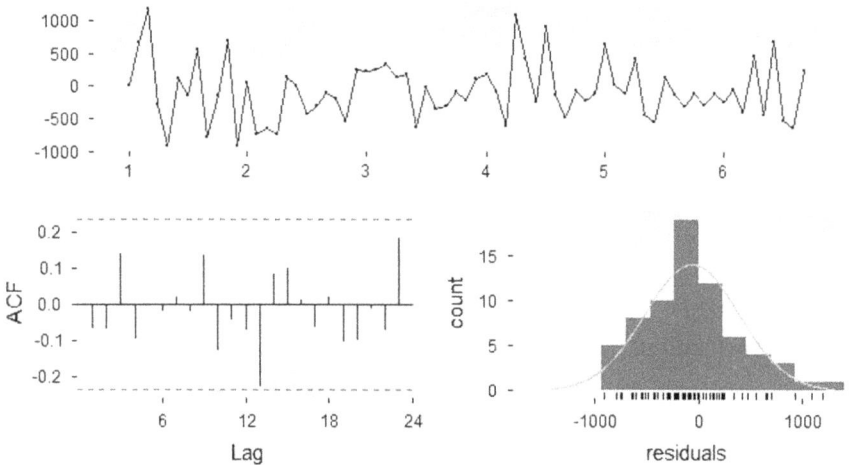

FIGURE 2.9 Residual analysis of ARIMAX model.

2.2.3.1   Simple Exponential Smoothing (SES)

Simple exponential smoothing (SES) is the simplest form of the exponential smoothing methods. If there is no apparent trend or seasonal fluctuations, then this SES method would be appropriate. Each data point is weighted using a parameter named as smoothing parameter, α, $(0 < \alpha \le 1)$. If the latest observation is denoted by y_t and the forecasted result at time $t+1$ is denoted by y_{t+1}, then the formula can be represented as:

$$y_{t+1} = \alpha y_t + \alpha(1 - \alpha)y_t + \ldots + \alpha(1 - \alpha)^{t-1}y_1 \tag{2.4}$$

The component form is as follows:

$$y_{t+1} = l_t \tag{2.5}$$

$$l_t = \alpha y_t + (1 - \alpha)l_{t-1} \tag{2.6}$$

2.2.3.2   Double Exponential Smoothing (DES)

If the time series represent a trend, then this DES method is appropriate. Apart from the smoothing parameter, there is an additional parameter β which is associated with the trend. The forecasted value (for the additive case) at k step-ahead is given as:

$$y_{t+1} = L_t + kT_t \tag{2.7}$$

where L_t is the level estimate and T_t is the trend estimate at time t
   L_t and T_t are updated using the following equations:

$$L_t = \alpha y_t + \alpha(1 - \alpha)(L_{t-1} + T_{t-1}) \tag{2.8}$$

$$T_t = \beta(L_t - L_{t-1}) + (1 - \beta)(T_{t-1}) \tag{2.9}$$

In the multiplicative trend the equations change as:

$$y_{t+1} = L_t * kT_t \tag{2.10}$$

$$L_t = \alpha y_t + \alpha(1 - \alpha)(L_{t-1} * T_{t-1}) \tag{2.11}$$

$$T_t = \beta(L_t/L_{t-1}) + (1 - \beta)(T_{t-1}) \tag{2.12}$$

2.2.3.3   Holt-Winters Seasonal Smoothing

If a time series exhibits both trend and seasonality, then Holt-Winters seasonal smoothing is appropriate. The Additive Holt-Winters model equation is:

$$y_{t+1} = L_t + kT_t + S_{t+k-m} \tag{2.13}$$

The level, trend and season components will update as:

$$L_t = \alpha(y_t - S_{t-m}) + (1 - \alpha)(L_{t-1} + T_{t-1}) \tag{2.14}$$

$$T_t = \beta(L_t - L_{t-1}) + (1 - \beta)(T_{t-1}) \tag{2.15}$$

$$S_t = \gamma(y_t - L_t) + (1 - \gamma)(S_{t-m}) \tag{2.16}$$

where α, β and γ are smoothing parameters to capture pattern, trend and seasonality, respectively.

In the case of multiplicative seasonality, the following equations will be used:

$$y_{t+1} = (L_t + kT_t)S_{t+k-m} \tag{2.17}$$

$$L_t = \alpha\left(\frac{y_t}{S_{t-m}}\right) + (1 - \alpha)(L_{t-1} + T_{t-1}) \tag{2.18}$$

$$T_t = \beta(L_t - L_{t-1}) + (1 - \beta)(T_{t-1}) \tag{2.19}$$

$$S_t = \gamma\left(y_t/L_t\right) + (1 - \gamma)(S_{t-m}) \tag{2.20}$$

Various models can be developed by changing additive or multiplicative structure for trend, seasonality and resulting error component.

2.2.3.4   Application of Exponential Smoothing Method

Monthly reported dengue data from January 2010 to December 2015 in Jakarta was used for model development. According to the ADF and KPSS tests, there was non-stationarity in the original series of Jakarta at a 5% significance level. Therefore, simple as well as double exponential smoothing techniques may not be appropriate for modeling the series. Hence, the Holt-Winters smoothing technique is more appropriate in predicting future dengue cases. We tested additive and multiplicative structures for trend, seasonality and error components of the series. We selected the most appropriate model which has minimum AIC, BIC, MAPE, MAE and RMSE measures as the best model to forecast future dengue cases in Jakarta. The best model consists of multiplicative error and multiplicative seasonality with no trend. Properties of the residuals were verified by applying the Ljung-Box test on residuals as well as on squared residuals (to validate the constant error variance assumption) at a 5% level of significance (corresponding p values were 0.8694 and 0.2892, respectively). According to the Jarque Bera test, residuals followed a normal distribution (p value = 0.6786) (Figure 2.10).

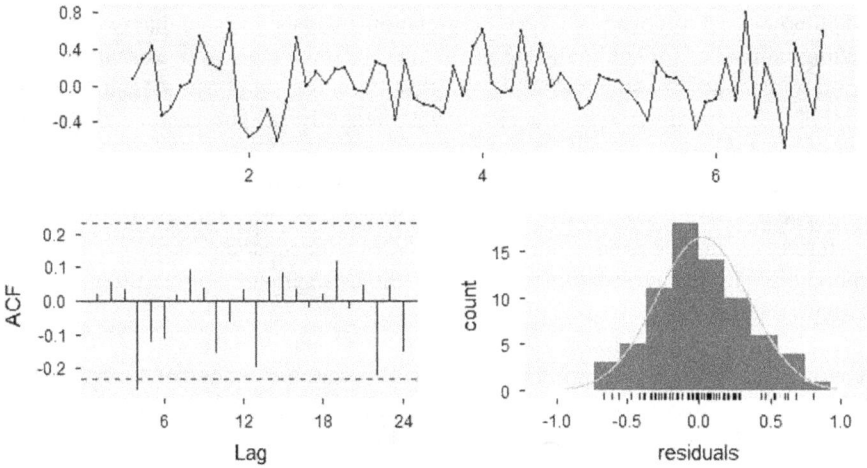

FIGURE 2.10   Residual analysis—exponential smoothing model.

The most appropriate model was validated using the data from January to March 2016, and the corresponding MAPE was less than 10%. Therefore, the selected best model is appropriate in short-term forecasting of dengue cases in Jakarta.

2.2.4   EXPONENTIAL SMOOTHING WITH EXPLANATORY VARIABLES (ETSX)

In order to improve the forecasting performances of the regular exponential smoothing method, covariates/independent variables or explanatory variables can be coupled with the method. The "es" function of the "smooth" package can be used for the purpose. There are two functions for additive and multiplicative error models. The formula for the former one is:

$$y_t = w'v_{t-1} + a_1 x_{1t} + a_2 x_{2t} + a_3 x_{3t} + \cdots + a_k x_{kt} + \varepsilon_t \qquad (2.21)$$

where a_1, a_2, ..., a_k are the parameters for independent variables x_{1t}, x_{2t}, ..., x_{kt}. ε_t is the error term which is assumed to be normally distributed. w' is the predefined measurement vector, and v_t is a state vector which consists of level, seasonal and trend components of the time series of interest.

The multiplicative error model is formulated as:

$$\log y_t = w' \log v_{t-1} + a_1 x_{1t} + a_2 x_{2t} + a_3 x_{3t} + \ldots + a_k x_{kt} + \log (1 + \varepsilon_t) \qquad (2.22)$$

2.2.4.1   Application of Exponential Smoothing with Explanatory Variables

The application is based on the monthly dengue cases reported in the Colombo district, Sri Lanka, from January 2010 to July 2019. We used the last three months' data to validate the model. We chose monthly rainfall, number of rainy days in a month, average wind speed and average monthly maximum temperature in the

Colombo district as covariates. Cross-correlation analysis revealed that two months leading rainfall and five months lag for number of rainy days were associated with monthly dengue cases (Table 2.6). Further, average wind speed and average maximum temperature were associated with monthly dengue cases with no lag periods (Table 2.6).

In addition, Pearson correlations of covariates with associated lag periods and dengue cases were statistically not significant at the 5% level of significance. Therefore, we modeled the monthly dengue cases with covariates using exponential smoothing with the explanatory variables method. A summary of the fitted models is shown in Table 2.7.

Table 2.7 shows the lower AIC, AICc and BIC values for covariates Rainfall, Average Wind Speed, Rainy Days and a two-variable model of Rainfall and Average Wind Speed. The normality assumption is only satisfied by the ETSX model with Rainy Days at a significance level of 5%. We verified other properties of the residuals of this model by applying the Ljung-Box test on residuals as well as on squared residuals at a 5% level of significance (corresponding p values were 0.9232 and 0.067, respectively). Therefore, we selected the ETSX model with number of Rainy Days as the covariate as the most appropriate model. We validated the most appropriate model using the data from May to July 2019, and the corresponding MAPE was 7.8%. MAPE values of the validation set are summarized in Table 2.7. Therefore, the selected best model is appropriate in short-term forecasting of dengue cases in Colombo, Sri Lanka. It was consistent with multiplicative error, multiplicative seasonal structures with no trend. The actual and fitted values of this model are shown in Figure 2.11.

TABLE 2.6
Cross-correlation of Variables

Lag in Months	Cross-Correlations			
	Rainfall, Dengue	Average Wind Speed, Dengue	Number of Rainy Days, Dengue	Maximum Temperature, Dengue
10	−0.036980	0.027344	0.053677	0.002296
9	−0.062257	−0.141262	0.094287	0.141674
8	−0.038958	−0.146240	0.097129	0.067203
7	−0.024077	0.037023	−0.088454	−0.070382
6	−0.051796	0.155418	−0.321836	−0.184028
5	−0.037607	0.009169	−0.369271	−0.170569
4	0.027498	−0.185926	−0.214652	−0.083364
3	0.155408	−0.251720	0.005366	−0.054397
2	0.318147	−0.091633	0.171329	−0.117476
1	0.288342	0.151775	0.169951	−0.169137
0	0.159166	0.255793	0.097682	−0.189566

TABLE 2.7

Summary of Fitted ETSX Models

Covariates in the Model	AIC	AICc	BIC	MAPE of Validation Set
Rainfall	1576.4	1582.5	1619.2	12
Rainy Days	1577.5	1583.5	1620.2	7.8
Average Wind Speed	1576.6	1582.7	1619.4	12
Maximum Temperature	1582.2	1589.9	1630.3	11.3
Rainfall, Average Wind Speed	1576.9	1583.8	1622.3	10
Average Wind Speed, Rainy Days	1578.3	1585.1	1623.7	10
Rainy Days, Maximum Temperature	1579.7	1586.5	1625.1	7.83
Rainfall, Rainy Days	1578	1584.9	1623.4	11.2
Average Wind Speed, Maximum Temperature	1578.7	1585.6	1624.1	11.8
Rainfall, Maximum Temperature	1578.3	1585.3	1623.8	11.9
Rainfall, Maximum Temperature, Average Wind Speed	1578.9	1586.7	1626.2	16
Rainfall, Maximum Temperature, Rainy Days	1579.9	1587.7	1628	11
Rainfall, Rainy Days, Average Wind Speed	1578.5	1586.2	1626.6	14
Maximum Temperature, Rainy Days, Average Wind Speed	1580.2	1588	1628.4	9.9
Maximum Temperature, Rainy Days, Average Wind Speed, Rainfall	1580.4	1589.2	1631.2	13.5

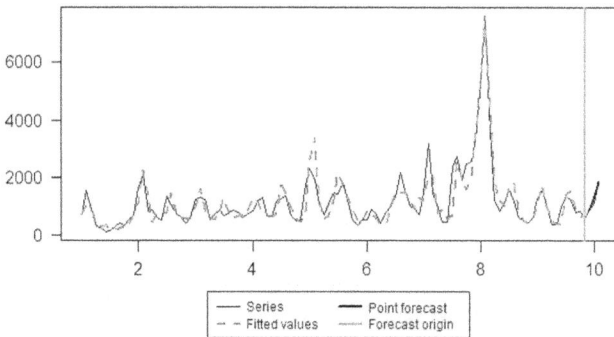

FIGURE 2.11 Actual and fitted value plot of the best ETSX model.

2.2.5 ALPHA-SUTTE MODELING

The Alpha-Sutte indicator method is applied to forecast data in various fields, including time series data. It originated from the Sutte indicator (Ahmar et al., 2017). The major advantage of this Alpha-Sutte modeling approach is that it requires only four observations to model and can predict the fifth (next) observation. Further, no assumptions exist to validate the model and forecasts. Therefore, this modeling approach is principally appropriate for short-term forecasting of a data series. The formula of the Alpha-Sutte method is shown below (Ahmar, 2018):

$$a_t = \frac{\alpha \left(\frac{\Delta x}{\frac{\alpha + \delta}{2}} \right) + \beta \left(\frac{\Delta y}{\frac{\alpha + \beta}{2}} \right) + \gamma \left(\frac{\Delta z}{\frac{\beta + \gamma}{2}} \right)}{3} \tag{2.23}$$

where $\delta = a_{t-4}$, $\alpha = a_{t-3}$, $\beta = a_{t-2}$, $\gamma = a_{t-1}$, $\Delta x = \alpha - \delta$, $\Delta y = \beta - \alpha$, $\Delta z = \gamma - \beta$

a_t is the observation at time t.

R programming language (R Core Team, 2020) is utilized to implement the functions and plots.

2.2.5.1 Applications of Alpha-Sutte Modeling

We used the cumulative number of monthly reported dengue cases in Jakarta, Indonesia, from the period of January 2010 to December 2015 for the analysis. We used monthly data in the year 2016 to validate the Alpha-Sutte method. The fitted values through the Alpha-Sutte approach and the actual dengue case values for the city are shown in Figure 2.12. It is apparent that the fitted values are following the actual data values very well.

The plot for the validation set is given in Figure 2.13. The calculated MAPE value for the validation set was 1.5%. As the percentage errors are less than 10% (Lewis, 1982), the Alpha-Sutte indicator method has higher accuracy in forecasting the cumulative monthly dengue cases Jakarta.

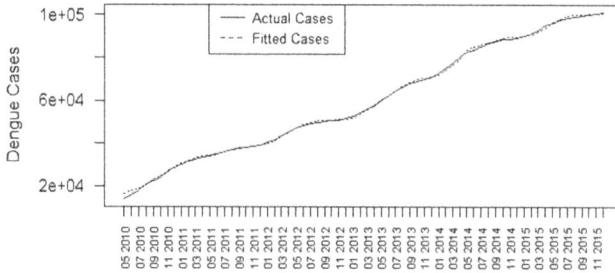

FIGURE 2.12   Actual and fitted value plot—Alpha-Sutte model.

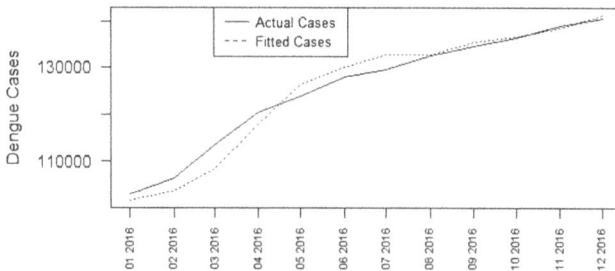

FIGURE 2.13   The validation plot—Alpha-Sutte model.

Further, we applied the method in modeling the cumulative number of monthly reported dengue cases in Colombo, Sri Lanka. The data period for the study was January 2010 to May 2020, and we used the last 10% of the data (from May 2019 to May 2020) to validate the Alpha-Sutte approach. The fitted model with actuals is shown in Figure 2.14. We can conclude from the figure that the Alpha-Sutte model was fitted on actual data values appropriately.

The validation results for the Colombo district are shown in Figure 2.15. The calculated MAPE value for the validated data was 0.69%. As the percentage error is less than 10% (Lewis, 1982), the Alpha-Sutte method has higher accuracy in forecasting the cumulative dengue cases in Colombo, Sri Lanka. The forecasts for the months of June to August 2020 were 146,534, 146,658 and 146,813 cumulative cases.

We can conclude from the analysis that the Alpha-Sutte approach performed well in modeling dengue cases in Jakarta as well as in Colombo. Therefore, the Alpha-Sutte method can be recommended for short-term forecasting of dengue cases in both cities.

2.2.6 TIME SERIES DECOMPOSITION

A time series is a combination of various patterns, and it is beneficial to divide the series into these patterns in order to understand the series as well as to enhance the accuracy of the forecasts. Generally, a time series includes trend (long-term variation in the series), seasonal (variation in the series at regular periods) and irregular/random (remainder after eliminating trend and seasonality of the series)

FIGURE 2.14   Actual and fitted value plot—Alpha-Sutte model.

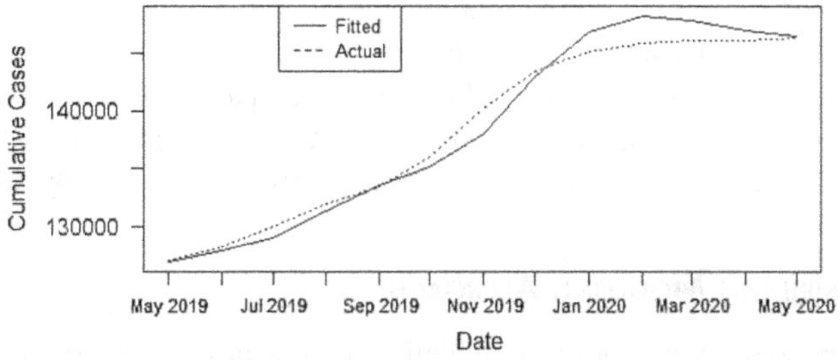

FIGURE 2.15   The validation plot—Alpha-Sutte model.

components. The time series decomposition involves separating the series into these components. There are two types of decompositions, namely additive and multiplicative. An additive decomposition is appropriate when the seasonal swing and trend components do not vary with the level of the series whereas when those change with the level of the series then multiplicative decomposition is appropriate. In the additive case, the series y_t can be written as:

$$y_t = S_t + T_t + R_t \qquad (2.24)$$

where S_t is the seasonal component, T_t is the trend component and R_t is the random component.

In the multiplicative case, the series can be written as:

$$y_t = S_t \cdot T_t \cdot R_t \qquad (2.25)$$

If the seasonal component is removed from the original series, then the resulting series is called seasonally adjusted data. The additive case is obtained by $y_t - S_t$ and the multiplicative y_t/S_t. Similarly, the de-trended series is obtained by $y_t - T_t$ and y_t/T_t for additive and multiplicative cases. The remainders are

$$R_t = y_t - S_t - T_t \qquad (2.26)$$

and

$$R_t = y_t/S_t \cdot T_t \qquad (2.27)$$

respectively.

The "decompose" function in the "stats" package is useful to decompose a time series into its components.

2.2.6.1 Application of Time Series Decomposition

The additive decomposed components (trend, seasonal and random) of the dengue series of Jakarta are shown in Figure 2.16. In the figure, the first panel shows the original series and then the trend component. The third panel of Figure 2.16 shows the seasonal component, and the final panel shows the random component of the series. If we add these final three plots together, it will produce the original series.

The multiplicative decomposed components are in Figure 2.17. Notice that the trend component is similar in Figure 2.16 and Figure 2.17. Apart from the graphical display of decomposed components, these components can be obtained through the "decompose" function and can be utilized to acquire de-trended and seasonally adjusted data.

2.2.7 COMBINING MODELING APPROACHES

Forecasts generated by two or more modeling approaches can be combined in order to improve the performances of models as well as to reduce bad performances

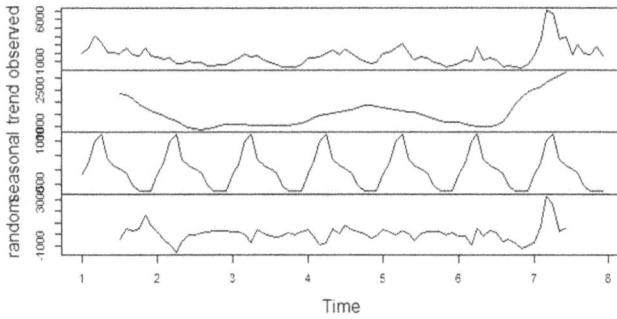

FIGURE 2.16   Additive decomposition.

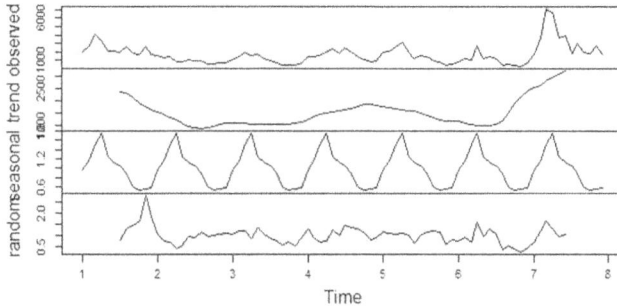

FIGURE 2.17   Multiplicative decomposition.

(Attanayake et al., 2019). Various combining methods are available in the literature (Christoph et al., 2018; Pantaznpoulos & Pappis, 1998). A list of combining approaches with summarized estimation procedure is given in Table 2.8.

2.2.7.1   Application of Combining Modeling Approaches

We used January 2010 to December 2015 data in Colombo for the development of models, and we used January 2016 to November 2017 data for validation. ADF and KPSS tests confirmed the non-stationarity in the series of Colombo. Non-stationarity was overcome by applying the first differencing. We tested various autoregressive and moving average parameters for the ARIMA model. We found the most appropriate model for Colombo that had minimum BIC, AIC and AICc measures. The selected model validates for assumptions. All the assumptions of residuals were satisfied by the optimum model at a 5% significance level. The model selected for Colombo was seasonal ARIMA $(1,1,2)$ $(1,0,0)_{12}$.

We applied Holt-Winters seasonal exponential smoothing for the same data series, and the method suits well because the series exhibits both seasonality and trend. We constructed different combinations of multiplicative and additive structures for all error, trend and seasonality components of the series. The most appropriate exponential smoothing model for Colombo consists of multiplicative

TABLE 2.8

Combining Forecast Methods

Combining Forecasting Method	Estimation Procedure (Combining Forecast is Given as f_c and p is Number of Approaches Used)
Simple Average	Arithmetic average of forecasts
Trimmed Mean	$f_c = \frac{1}{p(1-2\lambda)} \sum_{i=\lambda p+1}^{(1-\lambda)p} f_i$ (2.28); λ is the trimmed factor
Complete Subset Regression	$\sum_{i=1}^{p} \binom{p}{i}$
Standard Eigenvector Approach	$f_c = [\sum_{i=1}^{p} f_i w_i]$ (2.29); w_i is the weight $\sum_{i=1}^{p} w_i^2 = 1$
Bias-Corrected Eigenvector Approach	$f_c = \alpha + [\sum_{i=1}^{p} f_i w_i]$ (2.30); α corrects the bias.
Trimmed Bias-Corrected Eigenvector Approach	Bias-corrected Eigen vector approach is modified by removing poor models.
Inverse Ranking Approach	$f_c = \left[\sum_{i=1}^{p} f_i \frac{Rank_i^{-1}}{\sum_{j=1}^{p} Rank_j^{-1}} \right]$ (2.31) The model with the lowest mean squared error is assigned as Rank 1 and so on.
Median Approach	$f_c = f_{\left(\frac{p}{2}+0.5\right)}$; for odd p (2.32) $f_c = f_{\left(\frac{p}{2}\right)} + f_{\left(\frac{p}{2}+1\right)}$; for even p (2.33)
Newbold/Granger Approach	$f_c = F_{Np} \left[\frac{E^{-1}e}{e'E^{-1}e} \right]$ (2.34) e be a $p*1$ vector of $(1\ 1\ 1...1)'$ E be the mean squared prediction error matrix of F_{Np}.
Ordinary Least Squares (OLS) Regression	$f_c = \beta + \sum_{i=1}^{p} w_i f_i$ (2.35) OLS estimates estimated by minimizing the sum of squared errors in Inverse Ranking Approach
Bates/Granger Approach	$f_c = \left[\sum_{i=1}^{p} f_i \frac{\sigma^{-2}(i)}{\sum_{j=1}^{p} \sigma^{-2}(j)} \right]$ (2.36) $\sigma^2(i)$ is the estimated mean squared prediction error of model i
Winsorized Mean	$f_c = \frac{1}{p}[kf_{(k+1)} + \sum_{i=k+1}^{p-k} kf_{(p-k)}]$ (2.37); λ is the trimmed factor and $k = \lambda p$ (2.38)

error, multiplicative seasonality and additive structure for trend, and the selected model had the minimum AIC, BIC, MAE, MAPE and RMSE measures. Further, assumptions of residuals were satisfied by the model at a 5% significance level.

Validation results of exponential and ARIMA models are shown in Figure 2.18.

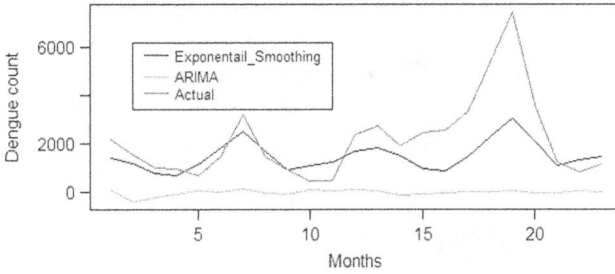

FIGURE 2.18   Forecasts of ARIMA and exponential smoothing models.

TABLE 2.9
Combining Forecasting Methods

Method	MAPE	RMSE
Simple Average	57.75	1993.69
Trimmed Mean	46.72	1620.77
Complete Subset Regression	55.54	988.16
Standard Eigenvector Approach	96.89	1225.74
Bias-Corrected Eigenvector Approach	62.83	1274.30
Trimmed Bias-Corrected Eigenvector Approach	62.82	1274.30
Inverse Ranking Approach	50.16	1774.85
Median Approach	57.76	1993.69
Newbold/Granger Approach	68.33	1074.91
Ordinary Least Squares (OLS) Regression	49.60	949.19
Bates/Granger Approach	46.72	1620.77
Winsorized Mean	57.75	1993.69

In order to improve the forecasts of ARIMA and exponential smoothing models, we tested various combining forecast methods. MAPE and RMSE of each of the combining forecast methods are depicted in Table 2.9.

Minimum RMSE value as 949.19 is recorded for OLS Regression method. The MAPE value for the same method is recorded as 49.6. The value is comparatively at a low level. Hence, we chose this method as the most suitable combining forecast method to predict monthly dengue cases in Colombo. Figure 2.19 shows the actual figures with the combined figure. By comparing Figure 2.18 and Figure 2.19, we concluded that the forecasted values were significantly improved by the combining method.

2.3   CONCLUSION AND DISCUSSION

This chapter deals with multiple time series modeling techniques, namely, Autoregressive Integrated Moving Average (ARIMA), exponential smoothing,

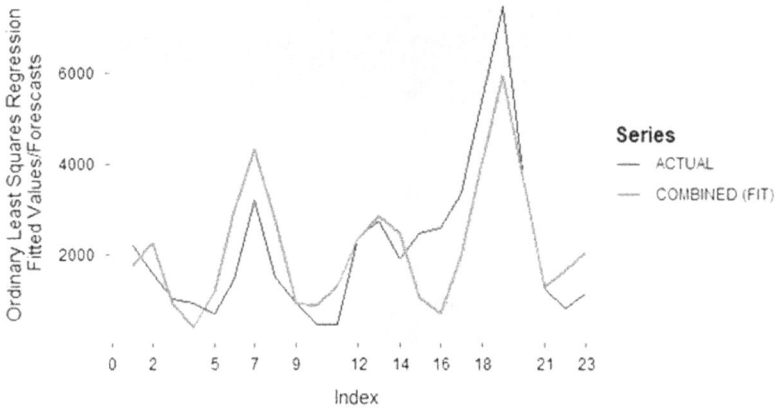

FIGURE 2.19 Combining forecast—OLS method.

decomposition, Alpha-Sutte modeling, Autoregressive Integrated Moving Average with Explanatory variables (ARIMAX) and exponential smoothing with explanatory variables. We illustrated the application of each technique using dengue cases reported in Jakarta and/or dengue cases reported in Colombo, Sri Lanka, as test cases. Monthly rainfall, number of rainy days within a month, average wind speed, and average maximum temperature data in Colombo were used as explanatory variables, whereas monthly rainfall and average humidity data in Jakarta were utilized for this purpose. Finally, we summarized several combined forecasting approaches and demonstrated how the forecasts of ARIMA and exponential smoothing techniques can be improved by combined forecasting approaches.

Different time series approaches have different advantages and disadvantages. The applicability of each technique depends on user requirements as well as the properties of the technique. The ARIMA technique is a widely used technique for time series analysis. The main advantage of this technique is it is a univariate technique; therefore, only the data of an interested variable is sufficient. Furthermore, it adapts with any frequency of data: daily, weekly, monthly or yearly, etc. The main disadvantage of the technique is that reliability of the selected model is highly dependant on the skills and experience of the modeler. Generally, the ARIMA technique is poor in predicting turning points and peaks and not appropriate in long-term forecasting. In addition, it is a challenging task to satisfy all the underlying assumptions. The ARIMA technique outlined in this chapter will provide a benchmark for other analysis.

If anyone wants to improve ARIMA modeling by incorporating external variables, then ARIMAX modeling would be an appropriate methodology. ARIMAX is a method that combines ARIMA and regression modeling; hence, it may express better performance than an individual ARIMA or regression approach. ARIMAX shows a relationship for a system of variables and ARIMA-type models require more than 50 observations to capture patterns existing in the data so that is one of the disadvantages of these categories of modeling.

Exponential smoothing is an alternative time series analysis method that performs well in short-term forecasting. It gives more weight to recent observations than past observations and hence uplifts future performances through the model. This technique may be inadequate in predicting turning points and peaks as well as selection based entirely on the experience and practice of the developer. In order to engage independent variables in exponential smoothing, one should consider ETSX modeling. Similar to ARIMAX, this method enhances the performances that exist in ETS modeling. However, the ETSX model building procedure is more complex than ETS modeling is.

Alpha-Sutte modeling is one of the newly invented time series modeling approaches. It is very simple to model, and only four observations are required to predict the fifth observation. Therefore, this method is an efficient method in forecasting future performance of the series. Furthermore, no assumptions exist to validate, which increases the applicability of the method. The two example test cases illustrated in this chapter performed well in modeling and predicting the dengue cases in Jakarta and Colombo by reporting lower MAPE values. Therefore, Alpha-Sutte modeling can be recommended for short-term forecasting of a series.

Time series decomposition provides a platform for time series analysis by investigating several components of a series. It is beneficial to divide the series into several components in order to understand the series as well as to enhance the accuracy of the forecasts. Although time series decomposition is a widely used method, we cannot recommend it as there are several drawbacks. One is that there are no results for the trend estimates for the first and the last few observations. Similarly, there is no estimate for the remainder/random component for the first and the last few observations. Furthermore, the trend estimates may over-smooth the sudden increases in the series. The seasonal component also assumes repeats every year, but in reality this may not true for longer series.

Combining forecasts is a useful strategy for improving forecasts generated by two or more techniques. Combining minimizes errors that may exist in individual techniques. The simple average of forecasting results may be a good starting point, and this chapter summarizes the variety of combining methods that are available in the field. This is a relatively new approach and can be used when different forecasting methods yield different results and the user is not sure which method to use. The example given in the chapter shows how much the forecasts generated by ARIMA and exponential smoothing can be improved by combining results. Therefore, the combined forecasting approach is an efficient strategy in the current modeling era over the classical approaches mentioned in the chapter.

The chapter outlines different time series approaches and which method is applicable for a given scenario depending on several factors. The main factor is user requirement. The user may need long-term forecasts or short-term forecasts, and precision of the results that he/she expects is further subjective. Another deciding factor is the size of available data. It is a challenging task to gather a large amount of data. Some analysis methods need more time in evaluating and checking assumptions. Therefore, time is also a deciding factor. The user may correspondingly consider the cost and benefits associated with the method as well as returns from the forecasts generated. One method may be well suited for a particular application, but the same method may not perform well in another application. Therefore, the decision of the most appropriate method is rather subjective and requires careful supervision, experience and investigation to select the best one.

Although one method is selected as the best-suited approach for a particular application, it is also subject to various uncertainties. Unexpected data recordings cannot be explained by fitted models. Uncontrolled and unknown factors may occur during the process. Nothing is certain in real-world scenarios; hence, the forecasts generated by the best approach may exhibit deviations from the actual one. Therefore, uncertainties continuously question the effectiveness and efficiency of a modeling approach.

REFERENCES

Ahmar, A. S. (2018). A comparison of α-Sutte indicator and ARIMA methods in renewable energy forecasting in Indonesia. *International Journal of Engineering & Technology*, *7*(1), 20–22.

Ahmar, A. S., Rahman, A., Arifin, A. N. M., & Ahmar, A. A. (2017). Predicting movement of stock of Y using sutte indicator. *Cogent Economics & Finance*, *5*(1). https://doi.org/10.1080/23322039.2017

Anggraeni, W., & Aristiani, L. (2016). Using Google Trend data in forecasting number of dengue fever cases with ARIMAX method case study: Surabaya, Indonesia, *International Conference on Information & Communication Technology and Systems (ICTS)*, 114–118. https://ieeexplore.ieee.org/document/7910283

Attanayake, A. M. C. H., Perera, S. S. N., & Liyanage, U. P. (2019). Combining forecasts of ARIMA and exponential smoothing models. *Advances and Applications in Statistics*, Pushpa Publishing House, Allahabad, India, *59*(2), 199–208. http://dx.doi.org/10.17654/AS059020199.

Attanayake, A. M. C. H., Perera, S. S. N., & Liyanage, U. P. (2020). Exponential smoothing on forecasting dengue cases in Colombo, Sri Lanka. *Journal of Science*, *11*(1), 11–22. http://doi.org/10.4038/jsc.v11i1.24

Box, G. E. P., & Jenkins, G. (1970). *Time series analysis: Forecasting and control*. Holden-Day.

Christoph, E. W., Eran, R., & Gernot, R. (2018). Forecast combinations in R using the ForecastComb package. *The R Journal*, *10*(2), 262–281.

Cortesa, F., Martellib, C. M. T., Ximenesa, R. A. A., Montarroyosa, U. S., Juniord, J. B. S., Cruze, O. G., Alexanderf, N., & Souza, W. V. (2018). Time series analysis of dengue surveillance data in two Brazilian cities. *Acta Tropica*, *182*, 190–197.

Dom, N. C., Hassan, A. A., Latif, Z. A., & Ismail, R. (2013). Generating temporal model using climate variables for the prediction of dengue cases in Subang Jaya, Malaysia. *Asian Pacific Journal of Tropical Disease*, *3*(5), 352–361.

Earnest, A., Tan, S. B., Smith, A. W., & Machin, D. (2012). Comparing statistical models to predict dengue fever notifications. *Computational and Mathematical Methods in Medicine*, 2012. http://dx.doi.org/10.1155/2012/758674

Eastin, M. D., Delmelle, E., Casas, I., Wexler, J., & Self, C. (2014). Intra-and inter seasonal autoregressive prediction of dengue outbreaks using local weather and regional climate for a tropical environment in Colombia. *The American Journal of Tropical Medicine and Hygiene*, *91*(3), 598–610.

Ekasari, R., Susanna, D., & Riskiyani, S. (2018). Climate factors and dengue fever in Jakarta 2011-2015. In *The 2nd international meeting of public health 2016, KnE life sciences*, 151–160. DOI: 10.18502/kls.v4i4.2273. Retrieved October 7, 2020, from https://www.researchgate.net/publication/326259422_Climate_Factors_and_Dengue_Fever_in_Jakarta_2011-2015.

Epidemiology Unit. (2020, October 1). Ministry of healthcare and nutrition, Sri Lanka, Dengue Update. http://www.epid.gov.lk

Ho, C. C., & Ting, C. Y. (2015). Time series analysis and forecasting of dengue using open data. In H. Badioze Zaman, P. Robinson, A. F. Smeaton, T. K. Shih, S. Velastin, A. Jaafar, & N. Mohamad Ali (Eds.), *Advances in visual informatics. IVIC 2015. Lecture notes in computer science* (p. 9429). Springer. https://doi.org/10.1007/978-3-319-25939-0_5

Lewis, C. D. (1982). *Industrial and business forecasting methods*. Butterworths Publishing.

Pantaznpoulos, S., & Pappis, C. (1998). New methods for combining forecasts. *Yugoslav Journal of Operations Research*, *8*(1), 103–109.

Polwiang, S. (2020). The time series seasonal patterns of dengue fever and associated weather variables in Bangkok (2003–2017). *BMC Infectious Diseases*, *20*(208), 10. https://doi.org/10.1186/s12879-020-4902-6

Promprou, S., Jaroensutasinee, M., & Jaroensutasinee, K. (2006). Forecasting dengue hae-morrhagic fever cases in Southern Thailand using ARIMA models. *Dengue Bulletin*, *30*, 99.

R Core Team (2020). *R: A language and environment for statistical computing*. R Foundation for Statistical Computing. https://www.R-project.org/.

Setiati, T. E., Wagenaar, J. F. P., Kruif, M. D., Mairuhu, A. T. A., Grop, E. C. M., & Soemantri, A. (2006). Changing epidemiology of dengue haemorrhagic fever in Indonesia.

Sirisena, P. D., & Noordeen, F. (2014). Evolution of dengue in Sri Lanka: Changes in the virus, vector and climate. *International Journal of Infectious Diseases*, *19*, 6–12.

Suneetha, T., Jayalakshmi, G., Kavitha, V., Mounika, S. V. N. S. G., Kavya, M., Lakshmi, G., Ramya, Madhavi, S., & Madhusha, P. (2018). Visualizing the future on the in-cidents of dengue fever in India by using the techniques of time series analysis. *International Journal of Emerging Technologies in Computational and Applied Sciences*, *25*(1), 6–12.

3 Time Series Analysis of COVID-19 Confirmed Cases in Select Countries

Anurag Barthwal[1] and Shwetank Avikal[2]
[1]Shiv Nadar Univesity, Greater Noida, India
[2]Graphic Era Hill University, Dehradun, India

3.1 INTRODUCTION

The world is facing a threat of a new kind this year, with the world economies shivering in the grip of the new pandemic—COVID-19. A pandemic is a disease that has spread over a large geographic region, covering several countries or continents and affecting a large percentage of people (Heath Kelly, 2011). While China, South Korea, Japan, Australia, New Zealand, the United States and European countries have recovered, with daily confirmed cases reduced to a few hundred, the situation in Brazil, Russia and India seems grim. This work presents a time series and probability distribution-based analysis of the pandemic.

3.1.1 EMERGENCE OF CORONAVIRUS DISEASE 2019 (COVID-19)

The epidemic Coronavirus disease 2019 (COVID-19) is an illness which was declared a pandemic by the World Health Organization (WHO). It is a contagious disease caused by severe acute respiratory syndrome coronavirus 2 (SARS-CoV-2) that severely affects a person's lungs and airways. The virus outbreak, similar to severe acute respiratory syndrome (SARS), was first identified in Wuhan, China, in 2019 and has spread worldwide to over 177 countries and territories (WHO, 2020).

It is a virus which was present in animals, possibly transmitted from animals to humans, and then spread from human to human. COVID-19 can vary from mild to severe illness, like pneumonia. The strains of virus are visible in two to 14 days after exposure. Major symptoms are fever, cough, shortness of breath and fatigue. People with poor immune strength and pre-existing conditions like heart disease, diabetes, lung infections and any nasal tract disease are more susceptible to the illness. The chances of death among elderly people are higher compared with the younger population (CDC USA, 2020; Zhang, 2020). The data provided by the Chinese Centre for Disease Control and Prevention (China CDC) reflect that fewer women die from

DOI: 10.1201/9781003102281-3

the virus than men. An analysis of statistics for 56,000 patients by the World Health Organization (WHO, 2020; Zhang, 2020) shows that 6% of patients become critical with lung and organ failure, septic shock and mortality risk; 14% experience serious symptoms such as difficulty breathing and shortness of breath and 80% develop minor symptoms like fever and cough, with some having pneumonia.

Tracing back the virus's footprint, common types of human coronavirus are 229E & NL63 (alpha coronavirus) and OC43 and HKU1 (beta coronavirus). Other human viruses have been transmitted from animals. The Middle East Respiratory Syndrome Coronavirus (MERS-CoV), which was transmitted to humans from dromedary camels in 2012, caused life-threatening illness with respiratory damage in a few cases (Chan & Chan, 2013). Severe Acute Respiratory Syndrome Coronavirus-1 (SARS-CoV-1) of 2002–2003, first emerged in Guangdong, southern China, and was transmitted to humans from civet cats. Approximately 8,000 people got infected from the syndrome, nearly killing about 774, whereas COVID-19 crossed this limit in two months (Matías-Guiu et al., 2020).

COVID-19 began spreading among commuters of air travel routes, as people unknowingly came across the earliest infected case internationally. Wuhan (China), Thailand and South Korea were the primary coronavirus "hot zones" which witnessed rampant increases in cases within weeks and eventually contaminated close contacts. Also, the health care workers got infected, and a few even died from the disease. Cases spread into multifold chains, broadening fast and producing series of newly infected persons, who became sick and died at a high percent. This mysterious epidemic left the countries in isolation and activated the public health care systems geographically. The propagation of the COVID-19 pandemic occurs in five stages (CDC USA, 2020; WHO, 2020):

i. *Phase 1:* The presence of a new virus which could have health implications to humans is reported in some part of the world.
ii. *Phase 2:* Cases of the infection of the virus in humans are reported, but there is no human-to-human transmission.
iii. *Phase 3:* Human-to-human transmission of the virus is confirmed.
iv. *Phase 4:* The propagation of the virus is accelerated, and containment measures like social distancing and lockdown are required.
v. *Phase 5:* The infection of the virus subsides after attainment of a peak.
vi. *Phase 6:* Even though the pandemic outbreak has diminished, precautions are taken as there is a possibility of a second wave. The research work related to statistical and time series analysis is discussed in the next section.

3.2 LITERATURE REVIEW

Statistical analysis of the COVID-19 time series is presented by a plethora of authors. Time series, probability distribution, pattern matching and fractal analysis approaches by selected authors are discussed in this section. Benvento et al. (2020) and Ceylan (2020) have presented an Autoregressive Integrated Moving Average (ARIMA) forecast model for prediction of fresh COVID-19 cases, with the help of epidemiological data provided by the Johns Hopkins University. Tanujit et al.

(2020) have used a hybrid ARIMA-wavelet model to forecast daily new COVID-19 cases, ten days ahead. Mohsen et al. (2020) have proposed the use of the two-piece scale-mixture-normal distributions auto-regressive (TP–SMN–AR) model to forecast the real-world new and recovered cases of COVID-19. Hongchao (2020) have investigated the relationship between daily COVID-19 cases with average temperature and relative humidity in 30 provinces of China, using the generalized additive model. The authors concluded that rise in temperature and humidity leads to fall in the daily new COVID-19 cases. Fang et al. (2020) have modeled the spread of COVID-19 and the impact of control measures using the Susceptible Exposed Infectious Recovered (SEIR) model. Effective reproductive number and data-fitting are used to predict peak confirmed cases.

Ndaïrou et al. (2020) proposed a compartment model for modeling the COVID-19 outbreak in Wuhan, China, with emphasis on super-spreaders. The sensitivity of the model is investigated with variation of parameters, and numerical simulations are performed to evaluate its suitability. Similarly Çakan (2020), Fredj and Chérif (2020) and Qianying et al. (2020) have proposed models based on the SEIR epidemic model and performed numerical simulation to validate their proposal. Rossa et al. (2020) used generalized logistic growth, sub-epidemic wave model and the Richards growth model to generate short-term forecasts of total cumulative COVID-19 cases in Chinese provinces.

As different trends in new confirmed COVID-19 cases are observed in different countries, and the time series has shown a depreciation after attaining a peak value, ARIMA and other time series forecast algorithms are not effective enough for long-time forecasting. This work investigates the shape and correlation between the COVID-19 new cases time series of different countries. Such a study is useful in characterizing the growth and propagation of the COVID-19 pandemic in a region.

3.3   COVID-19 TIME SERIES ANALYSIS

This work banks upon the experience of the pandemic in China, South Korea, Italy, France, Germany, the United Kingdom and Spain to characterize the spread in other regions of the world. The spread of the pandemic is studied by observing the number of fresh COVID-19 cases being reported in a region each day.

3.3.1   Fresh COVID-19 Cases

The number of new COVID-19 cases registered in China, South Korea, Italy, France, Germany, the United Kingdom (UK), Japan, Australia and Spain from January 2020 to May 31, 2020, is plotted in Figure 3.1 (Data Sources: ECDC, 2020; Ritchie et al., 2020).

The spread of the pandemic shows a gradual increase in the number of cases in the first phase, and after reaching a peak, it begins to fall. China reported more than 15,000 cases on February 15, 2020 (Figure 3.1) but the peak value of fresh cases in a day was less than 10,000 for other countries mentioned above. The number of COVID-19 fresh cases registered in China and South Korea from December 30, 2020 to May 31, 2020 is plotted in Figures 3.2 and 3.3.

The spread of the pandemic in the case of China and South Korea shows a gradual increase for the first 15 days, after which it began to fall. More than 15,000

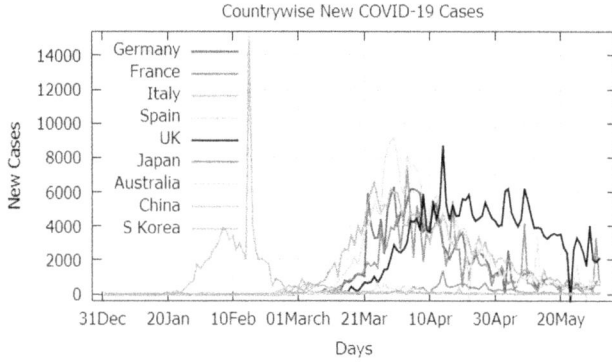

FIGURE 3.1   Fresh COVID-19 cases registered in China, South Korea, Japan, Australia, New Zealand, the United States and European countries from February 2, 2020 to May 31, 2020.

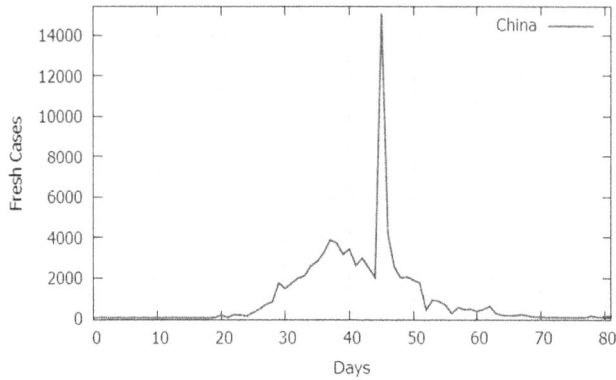

FIGURE 3.2   Fresh COVID-19 cases registered in China from December 31, 2020 to March 18, 2020.

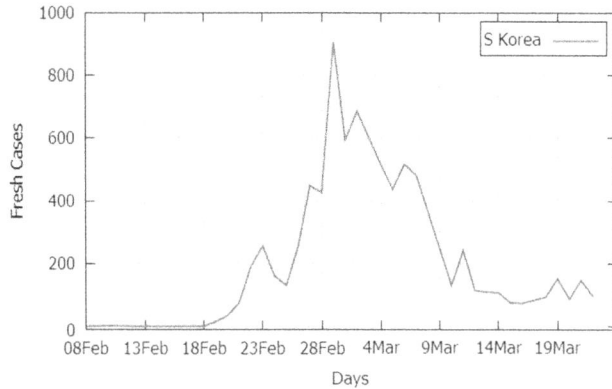

FIGURE 3.3   Fresh COVID-19 cases registered in South Korea from February 8, 2020 to March 18, 2020.

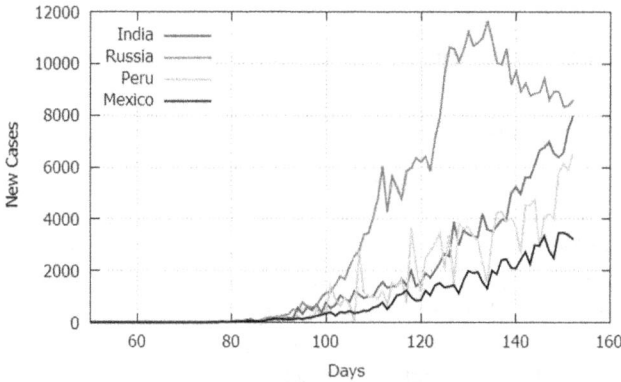

FIGURE 3.4 Fresh COVID-19 cases registered in India, Russia, Peru, India and Mexico from February 2, 2020 to May 31, 2020.

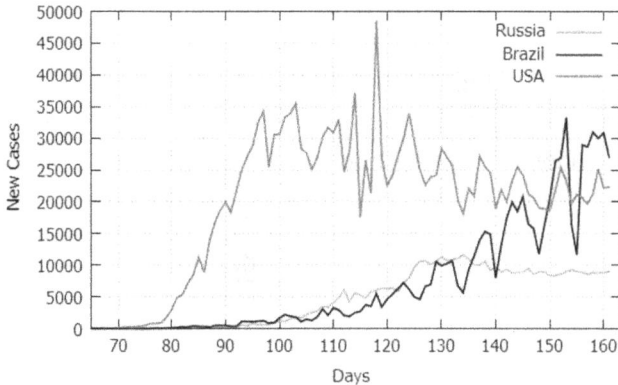

FIGURE 3.5 Fresh COVID-19 cases registered in Russia, Brazil and the United States from February 20, 2020 to May 31, 2020.

cases were reported on February 15, 2020 in China. After this spurt, the number of fresh cases began to fall again.

Unlike China, the fresh cases never crossed the 1,000 mark in South Korea, with the maximum cases reported on a single day (February 29, 2020) being 909. The 79-day observation saw an initial dormant phase, followed by a 10-day interval of persistent rise in fresh COVID-19 cases, followed by a steady decay phase. The propagation of the epidemic is described by way of plots of fresh cases being reported in other countries in Figures 3.4 and 3.5.

Figure 3.4 shows that Peru, India and Mexico are in the growth phase and are yet to attain a peak value. The peak has already arrived in Russia, with 11,656 cases on May 12, 2020, and the number of new COVID-19 cases has begun to decrease. Similarly, new cases in the United States have seen a peak on April 26, 2020 with

48,529 cases and are decreasing now (Figure 3.5). Brazil, on the other hand, is still showing an upward trend, with new cases increasing with each passing day.

3.4 METHODOLOGY

In this work, *cross-correlation coefficient* was used to establish a relationship between time series of new COVID-19 cases in different countries. The cross-correlation coefficient was used to determine if the relationship between time series of new cases was weak or strong, negative or positive. *Dynamic time warping* (DTW) was performed to estimate the similarity between the COVID-19 time series of the countries under observation. The time series data of those countries which witnessed a major COVID-19 outbreak, reached a daily peak value and recovered to a few hundred cases per day were considered for this study. The time series data of China, France, Italy, Germany, the United Kingdom and Spain were found to fulfill this criteria.

3.5 RESULTS AND DISCUSSION

3.5.1 CROSS-CORRELATION BETWEEN NEW CONFIRMED COVID-19 CASES OF DIFFERENT COUNTRIES

The cross-correlation between the time-series of new COVID-19 cases of different countries is calculated by using the formula:

$$r_k = \frac{\sum_{i=1}^{n-k}(X_i - \overline{X})(Y_{i+k} - \overline{Y})}{\sqrt{\sum_{i=1}^{n}(X_i - \bar{X})^2 \cdot \sum_{i-1}^{n}(Y_i - \overline{Y})^2}} \tag{3.1}$$

where $\overline{X}$ and $\overline{Y}$ are the mean values of the corresponding daily COVID-19 cases, respectively, and "k" is the time index. The correlation coefficient between time series of countries is obtained by comparing 25-day growth curves of new cases since 100 new cases were first reported.

3.5.1.1 Using the 25-Day Daily New COVID-19 Cases Since 100 New Cases Were First Reported (Growth Curve)

We studied the association between growth curves of new cases in different countries. The correlation between the 25-day growth curves of China, France, Italy, Germany, the United Kingdom and Spain were compared with India, which is still in the rising phase and has not attained a peak value. The time series of new COVID-19 cases for 25 days since 100 new cases were first reported is plotted in Figure 3.6.

The correlation coefficient between nation-pairs for new COVID-19 cases for 25 days since 100 new cases were first reported is presented in the form of a matrix in Table 3.1.

As is evident from Table 3.1, the new cases time series (growth phase) of Spain, Italy and the United Kingdom have the largest cross-correlation coefficient of 0.93, 0.91 and 0.91, respectively, with that of the new cases time series of India. It

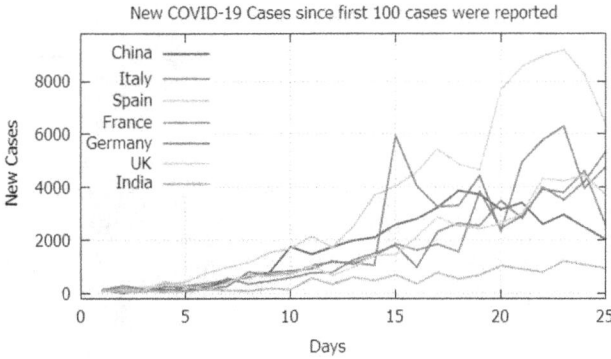

New COVID-19 Cases since first 100 cases were reported

FIGURE 3.6   New COVID-19 cases for 25 days since 100 new cases were first reported.

TABLE 3.1

Cross-Correlation Coefficients with New COVID-19 Cases for 25 Days Since 100 New Cases were First Reported

Country	China	Italy	Spain	France	Germany	United Kingdom	India
China	1	0.73	0.82	0.77	0.80	0.78	0.80
Italy	0.73	1	0.91	0.88	0.81	0.94	0.91
Spain	0.82	0.91	1	0.92	0.87	0.96	0.93
France	0.77	0.88	0.92	1	0.86	0.93	0.88
Germany	0.80	0.81	0.87	0.86	1	0.87	0.82
United Kingdom	0.78	0.94	0.96	0.93	0.87	1	0.91
India	0.80	0.91	0.93	0.88	0.82	0.91	1

---

indicates a strong and positive relationship between the new cases of these countries. The correlation with Germany and China is also strong and positive.

3.5.1.2   Using the New COVID-19 Cases Since 100 New Cases were First Reported to the Day when it Fell Back to Less than 100 Cases (Growth + Decay)

For the study, we considered countries like China, France, Italy, Germany, the United Kingdom and Spain, which witnessed a major COVID-19 outbreak, reached a daily peak value and recovered to a few hundred cases per day. The time series of new COVID-19 cases since 100 new cases were first reported to the day when it fell back to less than 100 cases (growth + decay) is shown in Figure 3.7. The correlation coefficient between nation-pairs for new COVID-19 cases for 25 days since 100 new cases were first reported is presented in the form of a matrix in Table 3.2.

As seen in Table 3.2, the new cases time series of Spain-Germany, Italy-United Kingdom and France-Germany show the highest degree of association, with a

New COVID-19 Cases since first 100 cases were reported

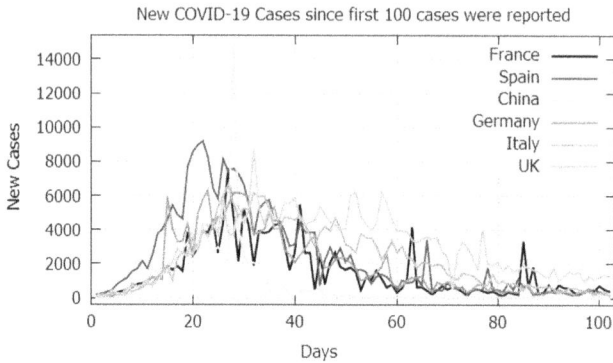

FIGURE 3.7   New COVID-19 cases since 100 new cases were first reported.

TABLE 3.2

Cross-Correlation Coefficients with New COVID-19 Cases For 25 Days Since 100 New Cases were First Reported

Country	France	Spain	China	Germany	Italy	United Kingdom
France	1	0.78	0.51	0.80	0.76	0.50
Spain	0.78	1	0.69	0.90	0.70	0.44
China	0.51	0.69	1	0.64	0.43	0.17
Germany	0.80	0.90	0.64	1	0.77	0.57
Italy	0.76	0.70	0.43	0.77	1	0.84
United Kingdom	0.50	0.44	0.17	0.57	0.84	1

---

correlation coefficient of 0.90, 0.84 and 0.80, respectively. This is an indication of a strong and positive association between new cases of these countries. The correlation with Spain-France, Italy-France and Italy-Spain is also strong and positive, with correlation coefficients of 0.78, 0.76 and 0.70, respectively.

3.5.2   DYNAMIC TIME WARPING

As observed in Figure 3.7, the growth rates (new COVID-19 cases) for the countries under study may be different, but they follow a similar trajectory, rising in the beginning and decaying after attaining a peak. To measure the similarity between such time series with similar shapes but different growth rates, it is suitable to use dynamic time warping (DTW).

DTW has been applied in the fields of audio, video, graphics, spoken word recognition, stock markets and power analysis. It computes an optimal match by calculating the Euclidian distance between each index of the first time series with one or more indices of the other. Let S and T be two time series to be matched:

$$S = s_0, \quad s_1, \quad s_2, \quad .. \ , s_N^N \tag{3.2}$$

and,

$$T = t_0, \quad t_1, \quad t_2, \quad .. \ , t_M^N \tag{3.3}$$

where n and m are the number of observations in time series S and T, respectively. The time series S and T are arranged in an n by m grid, with each grid point (i, j) corresponding to an alignment between the observation points s_i and t_j. The elements s_i and t_j are matched with the help of a warping function W so that the distance between them is minimum:

$$W = w_0, w_1, w_2, \quad .. \quad w_K^N \tag{3.4}$$

Thus, the warping path W is a sequence of grid points where each w_k corresponds to a point $(i, j)_K$. We then calculate the Euclidian distance between the elements of the two time series:

$$\delta(i, \ j) = \sqrt{(s_i - t_j)^2} \tag{3.5}$$

The DTW method is defined as a minimum of the potential warping paths, on the basis of the cumulative distance for each path:

$$DTW \ (S, \quad T) = min_W \left[\sum_{k=1}^{p} \delta(w_k) \right] \tag{3.6}$$

The normalized DTW distance is the difference between the maximum possible distance between time series S and T (M(S,T)) and $DTW (S, \ T)$ divided by M(S,T):

$$N(S, \ T) = \frac{M(S, \ T) - DTW(S, \ T)}{M(S, \ T)} \tag{3.7}$$

The normalized DTW distance metric obtained between different time series (Figure 3.7) is tabulated in Table 3.3.

Table 3.3 shows that the normalized DTW distance between Germany-Italy, France-Italy and France-Spain is the least, indicating that the new cases time series of these countries are the most similar.

The similarity between new cases time series for France, Germany, Italy and the United Kingdom (UK) is graphically represented in Figure 3.8 and Figure 3.9, respectively.

TABLE 3.3

DTW Distance Between Different Pair of Countries

Country	France	Spain	China	Germany	Italy	United Kingdom
France	0.0	286.68	322.41	316.25	283.95	423.37
Spain	286.68	0.0	494.05	329.99	272.00	408.15
China	322.41	494.05	0.0	345.98	313.92	642.77
Germany	316.25	329.99	345.98	0.0	220.86	397.0
Italy	283.95	272.00	313.92	220.86	0.0	300.87
United Kingdom	423.37	408.15	642.77	397.0	300.87	0.0

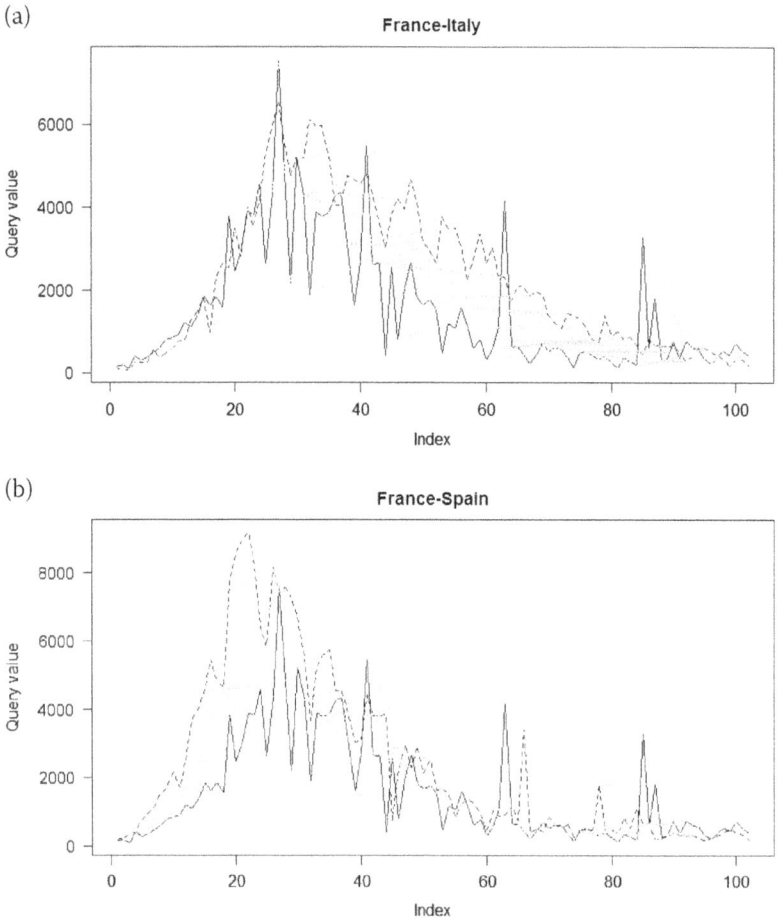

(a)

(b)

FIGURE 3.8 Computation of similarity between new cases time series of countries using DTW (France-Italy and France-Spain).

(a)

(b)

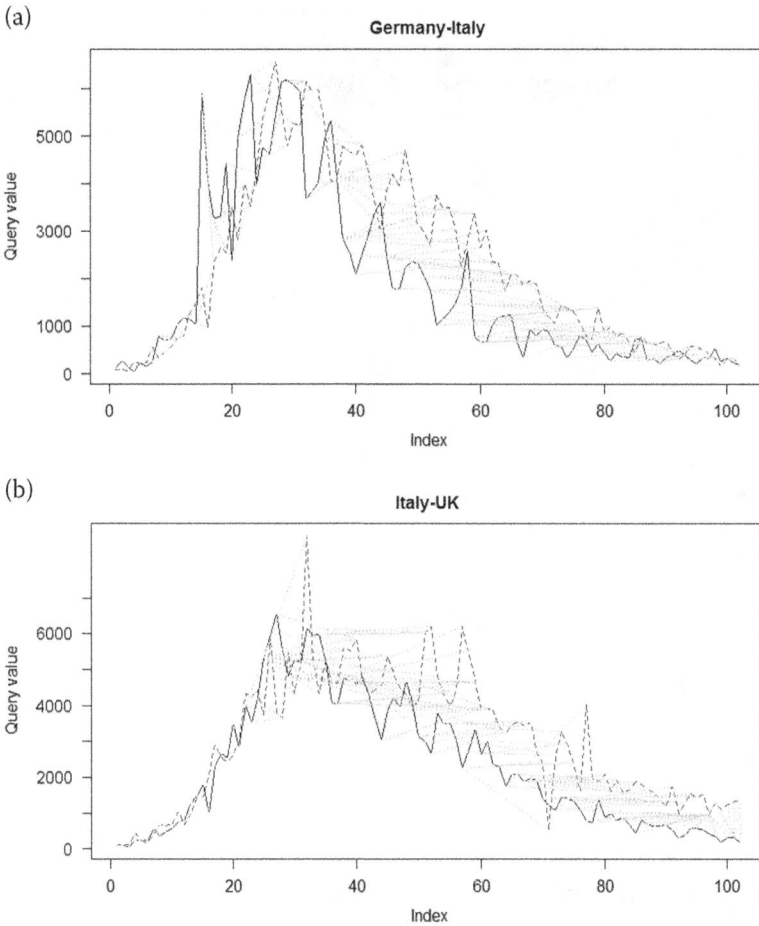

FIGURE 3.9 Computation of similarity between new cases time series of countries using DTW (Germany-Italy and Italy-United Kingdom [UK]).

On the other hand, the time series of China-United Kingdom and China-Spain are the most dissimilar. Table 3.4 shows the average normalized DTW (NDTW) distance between each of the five countries under study and the remaining countries:

France, Spain and Italy exhibit the least NDTW distance among themselves, indicating that the shapes of their new cases time series are most identical. The total COVID-19 cases of countries which are recovering from the pandemic and have eased lockdown restrictions show an S-shaped time series curve (Figure 3.10).

Here again, the trajectory of total COVID-19 cases is observed to follow a similar trajectory, demonstrating a steep rise in the beginning and flattening in the later stages. It shows that although the countries under study were different in terms of population size, demographics, government policies, health care system and implementation of containment policies, the trajectory of new cases in these countries followed a similar pattern.

TABLE 3.4
Average Normalized DTW Distance

Country	Avg. NDTW
France	326.532
Spain	358.174
China	423.826
Germany	322.016
Italy	278.32
United Kingdom	434.432

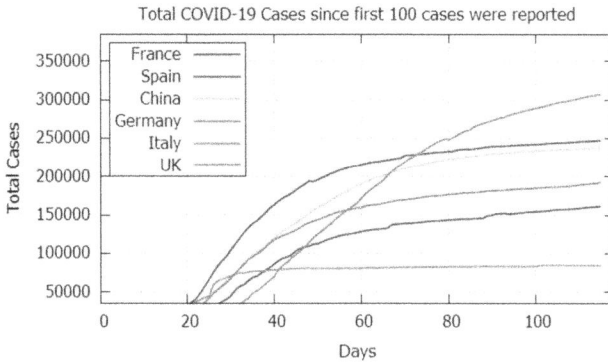

FIGURE 3.10   Total COVID-19 cases for countries under observation.

3.6   CONCLUSIONS

This chapter investigated the association between the time series of COVID-19 confirmed cases in select countries. We compared new cases growth curves, new cases time series and total cases time series of select countries with the help of correlation coefficients and DTW distance. We observed that there was a strong correlation between the time series of the countries under observation. The time series of France, Germany, Italy and the United Kingdom were found to be most similar. The strong correlation between the time series of France, Germany, Italy, the United Kingdom and China with that of India shows that the case study of these countries could be used to understand the propagation of the trajectory of COVID-19 cases in countries where the outbreak of the pandemic is in the initial stages. Social distancing, lockdown, personal hygiene and adoption of a healthy lifestyle could result in an early peak of fresh COVID-19 cases and help contain the pandemic with minimal losses of life and livelihood.

REFERENCES

Benvenuto, D., Giovanetti, M., Vassallo, L., Angeletti, S., & Ciccozzi, M. (2020). Application of the ARIMA model on the COVID-2019 epidemic dataset. *Data in Brief*, *29*. https://doi.org/10.1016/j.dib.2020.105340

Çakan, S. (2020). Dynamic analysis of a mathematical model with health care capacity for COVID-19 pandemic. *Chaos, Solitons, and Fractals*, 139. https://doi.org/10.1016/j.chaos.2020.110033

Ceylan, Z. (2020). Estimation of COVID-19 prevalence in Italy, Spain, and France. *Science of The Total Environment*, 729. https://doi.org/10.1016/j.scitotenv.2020.138817

Chakraborty, T., & Ghosh, I. (2020). Real-time forecasts and risk assessment of novel coronavirus (COVID-19) cases: A data-driven analysis. *Chaos, Solitons & Fractals*, *135*, 1–10. https://doi.org/10.1016/j.chaos.2020.109850

Chan, P. K. S., & Chan, M. C. W. (2013). Tracing the SARS-coronavirus. *Journal of Thoracic Disease*, *5*(2), S118–S121. https://doi.org/10.3978/j.issn.2072-1439.2013.06.19

Centers for disease control and prevention (CDA) USA (2020). Pandemic preparedness resources (2020, May 15). https://emergency.cdc.gov/planning/

Daily data on the geographic distribution of COVID-19 cases worldwide (2020). European Centre for Disease Prevention and Control (ECDC). https://www.ecdc.europa.eu/en/publications-data/download-todays-data-geographic-distribution-covid-19-cases-worldwide

Fang, Y., Nie, Y., Penny, M. (2020). Transmission dynamics of the COVID-19 outbreak and effectiveness of government interventions: A data-driven analysis. *Journal of Medical Virology*, *92*, 645–659. https://doi.org/10.1002/jmv.25750

Fredj, H. B., & Chérif, F. (2020). Novel Corona virus disease infection in Tunisia: Mathematical model and the impact of the quarantine strategy. *Chaos, Solitons, and Fractals*, *138*. https://doi.org/10.1016/j.chaos.2020.109969

Heath Kelly (2011). Bulletin of the World Health Organization. *World Health Organization (WHO)*, *89*, 540–541. http://dx.doi.org/10.2471/BLT.11.088815.

Hongchao, Q. (2020). COVID-19 transmission in Mainland China is associated with temperature and humidity: A time-series analysis. *Science of The Total Environment*, 728. https://doi.org/10.1016/j.scitotenv.2020.138778

Lin, Q. , Zhao, S., Gao, D., Lou, Y., Yang, S., Musa, S. S., Wang, M. H., Cai, Y., Wang, W., Yang, L., & He, D. (2020). A conceptual model for the coronavirus disease 2019 (COVID-19) outbreak in Wuhan, China with individual reaction and governmental action. *International Journal of Infectious Diseases*, *93*, 211–216. https://doi.org/10.1016/j.ijid.2020.02.058

Maleki, M., Mahmoudi, M. R., Wraith, D., & Pho, K.-H. (2020). Time series modelling to forecast the confirmed and recovered cases of COVID-19. *Travel Medicine and Infectious Disease*, *37*, 101742. https://doi.org/10.1016/j.tmaid.2020.101742.

Matías-Guiu, J., Gomez-Pinedo, U., Montero-Escribano, P., Gomez-Iglesias, P., Porta-Etessam, J., & Matias-Guiu, J. A. (2020). Should we expect neurological symptoms in the SARS-CoV-2 epidemic? *Neurologia*, *35*(3), 170–175. https://doi.org/10.1016/j.nrl.2020.03.001

Ndaïrou, F., Area, I., Nieto, J. J., Torres, D. F. M. (2020). Mathematical modeling of COVID-19 transmission dynamics with a case study of Wuhan. *Chaos, Solitons, and Fractals*, *135*, 1–6. https://doi.org/10.1016/j.chaos.2020.109846

Ritchie, H., Ortiz-Ospina, E., Roser, M., & Hasell, J. (2020, March 19). COVID-19 deaths and cases: How do sources compare?. *Our World in Data*. https://ourworldindata.org/covid-sources-comparison

Roosa, K., Lee, Y., Luo, R., Kirpich, A., Rothenberg, R., Hyman, J. M., Yan, P., Chowell, G. (2020). Real-time forecasts of the COVID-19 epidemic in China from February 5th to February 24th, 2020. *Infectious Disease Modelling*, *5*, 256–263.

World Health Organization (WHO) (2020) Rolling updates on coronavirus disease: COVID-19 (2020, May 30). https://www.who.int/emergencies/diseases/novel-coronavirus-2019

Zhang, Y. (2020, February 17). The epidemiological characteristics of an outbreak of 2019 novel coronavirus diseases (COVID-19). *China CCDC*. 10.46234/ccdcw2020.032. Available: http://weekly.chinacdc.cn/en/article/id/e53946e2-c6c4-41e9-9a9b-fea8db1a8f51

4 Bayesian Estimation of Bonferroni Curve and Zenga Curve in the Case of Dagum Distribution

Sangeeta Arora, Kalpana K. Mahajan, and Prerna Godura
Department of Statistics, Panjab University, Chandigarh-160014, India

4.1 INTRODUCTION

Dagum distribution, formulated by Camilo Dagum, helped to enable statistical distributions, which were used to fit empirical income and wealth data that could accommodate both heavy tails in empirical income and wealth distributions. One of the special cases of Dagum distribution actually appeared for the first time in 1942 (Burr, 1942) as the third example (Burr III) of solutions of the Burr distribution system. Since the Dagum distribution is the reciprocal transformation of the Burr XII, it is also called the inverse Burr, especially in cases of actuarial literature.

Since Dagum proposed his model as income distribution, its properties have been appreciated in economics and financial fields, and its features have been extensively discussed in the studies of income and wealth. Kleiber and Kotz (2003) provided a review on the origin of the Dagum model and its applications. Dagum distribution consists of both Type I and Type II specifications, where Type I is the three-parameter specification and Type II deals with the four-parameter specification. A detailed description of this distribution can be found in "A Guide to the Dagum Distribution" (Dagum, 1983).

A continuous random variable X is said to have a three-parameter Dagum distribution, abbreviated as $X \sim Dag\,(\alpha, \beta, q)$, if its cumulative distribution and probability density functions are given as

$$F(x) = \left[1 + \left(\frac{x}{\beta}\right)^{\alpha}\right]^{-(q+1)}, x > 0, \tag{4.1}$$

$$f_D(x) = \frac{\alpha q}{\beta^{\alpha q}} x^{\alpha q - 1}\left[1 + \left(\frac{x}{\beta}\right)^{\alpha}\right]^{-(q+1)}, x > 0, (\alpha, \beta, q > 0), \tag{4.2}$$

DOI: 10.1201/9781003102281-4

51

where α and q are shape parameters and β is a scale parameter.

For $q = 1$, Dagum distribution was also referred to as log-logistic distribution (Dagum, 1975).

The corresponding likelihood function of the Dagum distribution is given by

$$L = \left(\frac{\alpha q}{\beta^{\alpha q}}\right)^n \prod_{i=1}^{n} x_i^{\alpha q-1} \prod_{i=1}^{n} \left[1 + \left(\frac{x_i}{\beta}\right)^{\alpha}\right]^{-(q+1)}, \forall \ x > 0, (\alpha, \beta, q > 0) \quad (4.3)$$

It can be seen from expression (4.2) that Dagum distribution is a mixture of Generalized Gamma and Inverse Weibull distribution (Ahmed et al., 2017). The parameters of Dagum distribution have been estimated with censored samples by Domma (2007). Domma et al. (2011)estimated the parameters of the right truncated Dagum distribution by using maximum likelihood estimation. The estimation and skewness test for Dagum distribution were studied by Garvey et al. (2002).

In the context of Bayesian set-up, however, some results are reported in the case of Dagum distribution for the Lorenz curve, the classical inequality curve along with its prior and posterior distribution (Chotikapanich & Griffiths, 2006). However, much work needs to be done for other inequality measures: the Bonferroni and Zenga curves. Bayesian estimation for the Bonferroni curve is derived by Giorgi and Crescenzi (2001), and Arora et al. (2014) derived it for the Zenga curve in the case of the Pareto distribution under informative and conjugate priors. Some results for the Lorenz curve and associated Gini index in the Bayesian framework are available for exponential and generalised inverted exponential distribution (Mahmoud et al., 2017; Sathar et al., 2005), but for Dagum distribution, Bayesian estimators for inequality measures still await the attention of researchers.

In this chapter, our focus is on the derivation of Bayesian estimators for two main inequality curves—Bonferroni curve and Zenga curve—using both informative and non-informative priors along with different choices of loss functions.

The concept of inequality curves and their derivations in the case of Dagum distribution are given in Section 4.2. Section 4.3 deals with the Bayesian estimation of the Bonferroni curve under different priors and different loss functions. The simulation study using R software has been carried out in Section 4.4 to compare the efficiency of different Bayesian estimators on the basis of different estimated loss functions using different configuration of sample sizes and parameters. Section 4.5 gives the expressions for Bayesian estimation of Zenga curves under different priors and loss functions, and the simulation work is carried out in Section 4.6. In Section 4.7, a real-life example is given, along with the simulations of various results derived in previous sections.

4.2   MEASURING INEQUALITY CURVES USING DAGUM DISTRIBUTION

Inequality is a vital characteristic of non-negative distributions. It is used to analyze data in socio-economic sciences, in context to income distribution. The Lorenz curve (Lorenz, 1905) is the most used and oldest curve. Another classical curve related to the Lorenz curve is the Bonferroni curve (Bonferroni, 1930). However, in recent years, a new curve—Zenga curve—was introduced by Zenga (2007), which although being related to the other two curves, can assume different shapes, which allows it to distinguish different situations in terms of inequality.

These three curves have one characteristic in common, that they can be defined using only the mean of the whole population and the means of particular sub-groups. An important application of these inequality curves is that they can be used to define some orderings (Atkinson, 1970; Muliere & Scarsini, 1989). Such orderings allow the comparison of distributions in terms of inequality.

For continuous and non-negative random variable X, with the cumulative distribution function $F(x)$, the Lorenz, Bonferroni and Zenga curves are respectively defined as

$$L(u) = \frac{\int_0^u t f(t)dt}{E(x)}, \ 0 < u < 1,$$

$$B(u) = \frac{\int_0^u t f(t)dt}{F(x)E(x)}, \ 0 < u < 1,$$

$$Z(u) = 1 - \frac{\mu^-(u)}{\mu^+(u)}, \ 0 < u < 1,$$

where $E(x)$ is the mean income, $\mu^-(u) = \frac{\int_0^u t f(t)dt}{F(x)}$ and $\mu^+(u) = \frac{\int_0^u t f(t)dt}{1 - F(x)}$ are the lower and upper means, respectively.

The Lorenz curve gives the connection between the cumulative proportion of income units and the cumulative proportion of income received when the incomes are arranged in ascending order. The Lorenz curve has an intuitive appeal, and its graphical representation is an added advantage. The Lorenz curve, $L(u)$, itself is a distribution function with support on [0, 1] (Aaberge, 2000; Klefsjö, 1984; Kleiber, 2003).

The Bonferroni curve represents inequality in an equivalent manner to the Lorenz curve, and both curves are determined mutually; however, the information they yield is different. The values of L(u) are fractions of the total income, while the values of B(u) refer to relative income levels. These two measures have some applications in reliability and life testing as well (cf. Giorgi & Crescenzi, 2001). The Bonferroni curve can be represented graphically in the unit square and can also be related to the Lorenz curve (Figure 4.1)

The Zenga curve (Zenga, 2007) is a more recent curve of inequality. The main difference between the Lorenz curve and the Zenga curve is that the Zenga curve compares adjacent and disjoint parts of distribution, but the Lorenz curve makes the comparison of inequality based on cumulative, ordered and relative values.

The value of the Zenga curve (p) is between 0 and 1 (Figure 4.2). The Zenga curve compares the ratio of the mean income of the $p \times 100\%$ poorest to the $(1 - p) \times 100\%$ richest part of the population, thus giving an idea of inequality.

For Dagum distribution, these inequality curves are given by

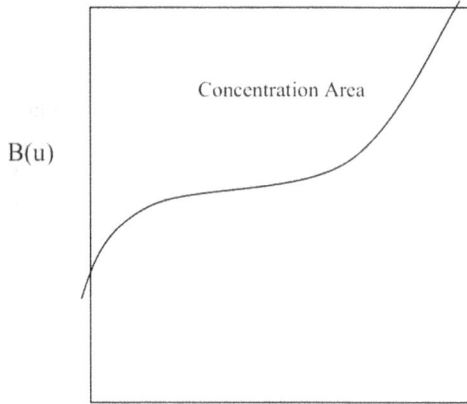

FIGURE 4.1 Bonferroni curve.

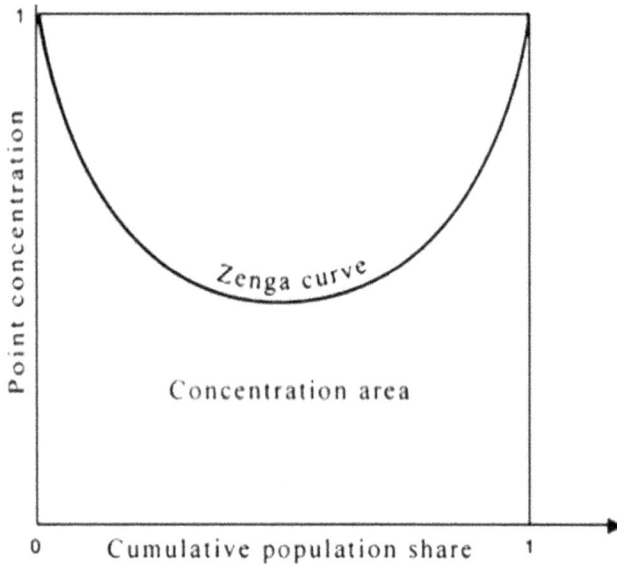

FIGURE 4.2 Zenga curve.

$$\text{Lorenz curve,} \quad L_D(u) = B_w\left[\left(q + \frac{1}{\alpha}\right),\left(1 - \frac{1}{\alpha}\right)\right], 0 < u < 1, (\alpha, q > 0), \quad (4.4)$$

where $w = u^{1/q}$, $0 < u < 1$, and $B_w(c, d)$ is the incomplete Beta integral defined as

$$B_w(c, d) = \frac{\int_0^w t^{c-1}(1 - t)^{d-1}dt}{\int_0^1 t^{c-1}(1 - t)^{d-1}dt}.$$

Bonferroni curve $B_D(u)$ is given by

$$B_D(u) = \frac{1}{q} B_w\left[\left(q + \frac{1}{\alpha}\right), \left(1 - \frac{1}{\alpha}\right)\right], \quad 0 < u < 1, \, (\alpha, \, q > 0), \qquad (4.5)$$

and

$$\text{Zenga curve } Z_D(u) = \frac{q - B_w\left[\left(q + \frac{1}{\alpha}\right), \left(1 - \frac{1}{\alpha}\right)\right]}{q\left[1 - B_w\left[\left(q + \frac{1}{\alpha}\right), \left(1 - \frac{1}{\alpha}\right)\right]\right]}, \quad 0 < u < 1, \, (\alpha, \, q > 0).$$

$$(4.6)$$

It can be seen from these equations that all three curves depend on the shape parameters q and α only.

4.3 BAYESIAN ESTIMATOR OF POINT MEASURE OF BONFERRONI CURVE $B_D(u)$, $0 < u < 1$ UNDER DIFFERENT PRIORS USING DIFFERENT LOSS FUNCTIONS

The choice of the loss function plays a very important role in Bayesian analysis. In this section, we use squared error loss function (symmetric loss function) and entropy loss function (asymmetric loss function) under informative and non-informative priors to obtain the Bayes estimator of the point measure of the Bonferroni curve $B_D(u)$, $0 < u < 1$.

Squared Error Loss Function (SELF)
A very simple and common symmetric loss function used in literature is the squared error loss function (SELF), a type of loss function which increases quadratically with the difference

$$L(\theta, \hat{\theta}) = (\theta - \hat{\theta})^2.$$

The estimated loss of parameter θ using SELF is obtained by

$$EL_S(L(\theta, \hat{\theta}) = \int_0^\infty (\theta - \hat{\theta})^2 {}_* \pi(\theta|x) d\theta, \qquad (4.7)$$

where $\pi(\theta|x)$ is the posterior density of the parameter θ.

It is seen that usually SELF is used for Bayesian inference when equal importance is given to the losses due to over- and under-estimation. However, in many situations over-estimation is more serious than under-estimation and vice versa. In order to deal with such situations, an appropriate and flexible class of asymmetric loss functions have been introduced in the literature. One such loss function is the entropy loss function.

Entropy Loss Function (ELF)
Calabria and Pulcini (1990) introduced a useful asymmetric loss function called the entropy loss function (ELF)

$$L(\theta, \hat{\theta}) = \delta\left[\left(\frac{\hat{\theta}}{\theta}\right) - ln\left(\frac{\hat{\theta}}{\theta}\right) - 1\right],$$

where the minimum occurs at $\hat{\theta} = \theta$.

Dey et al. (1987) studied this loss function for simultaneous estimation of scale parameters and their reciprocals in the case of gamma distribution.

The Bayes estimator of parameter θ in this case is

$$\hat{B}_E(\theta) = [E(\theta^{-1}|x)]^{-1}, \tag{4.8}$$

where $E(\theta^{-1}|x) = \int_0^\infty \frac{1}{\theta}*\pi(\theta|x)d\theta$.

The estimated loss for parameter θ using ELF is given by

$$EL_E(L(\theta, \hat{\theta}) = \int_0^\infty L(\theta, \hat{\theta})*\pi(\theta|x)d\theta, \tag{4.9}$$

where $\pi(\theta|x)$ is the posterior density of the parameter θ.

Choice of Prior
To obtain the Bayesian estimator of point measure of the Bonferroni curve $B_D(u)$, $0 < u < 1$, we use one informative prior—Mukherjee-Islam (MI)—and two non-informative priors—uniform and quasi prior.

Mukherjee-Islam Prior
Mukheerji and Islam (1983) is a well-rooted probability distribution that has been used by many researchers as a failure distribution for the purpose of reliability and Bayesian analysis.

The Mukherjee-Islam prior for the parameter θ is given by

$$\pi_M(\theta) = \rho\sigma^{-\rho}\theta^{\rho-1}, \quad \rho > 0, \sigma > 0, \theta > 0.$$

Uniform Prior
An obvious choice for non-informative prior is the uniform prior. The uniform prior for parameter θ is defined as

$$\pi_U(\theta) \propto 1, 0 < \theta < \infty.$$

A uniform (or flat) prior is one that doesn't favor any particular value of θ over another.

Quasi Prior

When there is no more prior information about the distribution of parameter θ, one may use the quasi prior density given by

$$\pi_Q(\theta) = \frac{1}{\theta^d}, \; \theta > 0, \; d > 0.$$

4.3.1 BAYESIAN ESTIMATION OF POINT MEASURE OF BONFERRONI CURVE $B_D(u)$, $0 < u < 1$, UNDER MUKHERJEE-ISLAM PRIOR

Case I: Shape parameter q is unknown and α, β are known

Assume the Dagum distribution has Mukherjee-Islam prior with hyper-parameters (ρ_1, σ_1); then, the density for q is given as

$$\pi_M(q) = \rho_1 \sigma_1^{-\rho_1} q^{\rho_1-1}, \; q > 0, \; \rho_1 > 0, \; \sigma_1 > 0. \tag{4.10}$$

The likelihood function (4.3) when shape parameter α and scale parameter β are known as

$$L(\alpha, \beta, q) \propto q^n \prod_{i=1}^{n} \left(\frac{x_i}{\beta}\right)^{\alpha q} \prod_{i=1}^{n} \left(1 + \left(\frac{x_i}{\beta}\right)^{\alpha}\right)^{-q}$$

$$= q^n e^{q \Sigma \ln \left(\frac{x_i}{\beta}\right)^{\alpha}} e^{-q \Sigma \log \left(1 + \left(\frac{x_i}{\beta}\right)^{\alpha}\right)}$$

$$\propto q^n e^{-q \Sigma_{i=1}^{n} \log \left(1 + \left(\frac{x_i}{\beta}\right)^{\alpha}\right)} = q^n e^{-qT}$$

where $T = \Sigma_{i=1}^{n} \log \left(1 + \left(\frac{x_i}{\beta}\right)^{\alpha}\right)$.

The posterior distribution of q is derived as

$$\pi_M^*(q) = \frac{L_* \pi_M(q)}{\int_0^\infty L_* \pi_M(q) dq}$$

$$= \frac{e^{-qT} q^{n+\rho_1-1}}{\int_0^\infty e^{-qT} q^{n+\rho_1-1} dq} \tag{4.11}$$

$$\pi_M^*(q) = \frac{T^{n+\rho_1} e^{-qT} q^{n+\rho_1-1}}{\Gamma(n+\rho_1)}.$$

The Bayes estimator of the Bonferroni curve $B_D(u)$ at some specific point u where $0 < u < 1$, in the case of **Mukherjee-Islam prior using SELF** is derived as

$$\hat{B}_{SM-q}(u) = \int_0^\infty B_D(u) * \pi_M^*(q|x) dq, \ 0 < u < 1$$

$$= \int_0^\infty \frac{1}{q} B_w \left[\left(q + \frac{1}{\alpha} \right), \left(1 - \frac{1}{\alpha} \right) \right] * \frac{T^{n+\rho_1} e^{-qT} q^{n+\rho_1-1}}{\Gamma(n+\rho_1)} dq$$

$$\hat{B}_{SM-q}(u) = \frac{T^{n+\rho_1}}{\Gamma(n+\rho_1)} \int_0^\infty B_w \left[\left(q + \frac{1}{\alpha} \right), \left(1 - \frac{1}{\alpha} \right) \right] * q^{n+\rho_1-1} e^{-qT} dq, \ 0 < u < 1.$$

$$(4.12)$$

Similarly, under the **ELF,** the Bayes estimator of the point measure $B_D(u)$, $0 < u < 1$, using **Mukherjee-Islam prior** is given as

$$\hat{B}_{EM-q}(u) = [E(B^{-1}|x)]^{-1}, \ 0 < u < 1,$$

$$E(B^{-1}|x) = \int_0^\infty \frac{1}{B_D(u)} * \pi_M^*(q|x) dq, \ 0 < u < 1$$

$$= \left[\frac{T^{n+\rho_1}}{\Gamma(n+\rho_1)} \int_0^\infty B_w \left[\left(q + \frac{1}{\alpha} \right), \left(1 - \frac{1}{\alpha} \right) \right]^{-1} * q^{n+\rho_1} * e^{-qT} d \right]^{-1}.$$

Therefore,

$$\hat{B}_{EM-q}(u) = \left[\frac{T^{n+\rho_1}}{\Gamma(n+\rho_1)} \int_0^\infty B_w \left[\left(q + \frac{1}{\alpha} \right), \left(1 - \frac{1}{\alpha} \right) \right]^{-1} * q^{n+\rho_1} * e^{-qT} dq \right]^{-1},$$

$$0 < u < 1.$$

$$(4.13)$$

Case II: Shape parameter α is unknown and q, β are known

The density of α under Mukherjee-Islam prior with hyper-parameters (ρ_2, σ_2) is given as

$$\pi_M(\alpha) = \rho_2 \sigma_2^{-\rho_2} \alpha^{\rho_2-1}, \ \alpha > 0, \rho_2 > 0, \sigma_2 > 0, \qquad (4.14)$$

where α_2 and σ_2 are the shape and scale parameters.

The posterior distribution of α is

$$\pi_M^*(\alpha) = \frac{L * \pi_M(\alpha)}{\int_0^\infty L * \pi_M(\alpha) da}$$

$$= \frac{\left(\frac{1}{\beta^{\alpha q}} \right)^n \Pi_{i=1}^n x_i^{\alpha q-1} \left[1 + \left(\frac{x_i}{\beta} \right)^\alpha \right]^{-(q+1)} \alpha^{n+\rho_2-1}}{\int_0^\infty \left(\frac{1}{\beta^{\alpha q}} \right)^n \Pi_{i=1}^n x_i^{\alpha q-1} \left[1 + \left(\frac{x_i}{\beta} \right)^\alpha \right]^{-(q+1)} \alpha^{n+\rho_2-1} d\alpha}. \qquad (4.15)$$

The Bayes estimator in the case of **Mukherjee-Islam prior** for the point measure Bonferroni curve $B_D(u)$, $0 < u < 1$, under **SELF** is obtained as

$$\hat{B}_{SM-\alpha}(u) = \int_0^\infty B_D(u) *\pi_M^*(\alpha|x)d\alpha, \ 0 < u < 1$$

$$= \int_0^\infty \frac{1}{q} B_w\left[\left(q+\frac{1}{\alpha}\right), \left(1-\frac{1}{\alpha}\right)\right] * \frac{\left(\frac{1}{\beta^{\alpha q}}\right)^n \Pi_{i=1}^n x_i^{\alpha q-1}\left[1+\left(\frac{x_i}{\beta}\right)^\alpha\right]^{-(q+1)} \alpha^{n+\rho_2-1}}{\int_0^\infty \left(\frac{1}{\beta^{\alpha q}}\right)^n \Pi_{i=1}^n x_i^{\alpha q-1}\left[1+\left(\frac{x_i}{\beta}\right)^\alpha\right]^{-(q+1)} \alpha^{n+\rho_2-1}d\alpha} \quad (4.16)$$

Similarly, under the **ELF**, the Bayes estimator of the point measure $B_D(u)$, $0 < u < 1$, using **Mukherjee-Islam prior** is given as

$$\hat{B}_{EM-\alpha}(u) = [E(B^{-1}|x)]^{-1}, \ 0 < u < 1$$

$$E(B^{-1}|x) = \int_0^\infty \frac{1}{B_D(u)} *\pi_M^*(\alpha|x)d\alpha, \ 0 < u < 1$$

$$= \int_0^\infty \frac{1}{\frac{1}{q}B_w\left[\left(q+\frac{1}{\alpha}\right),\left(1-\frac{1}{\alpha}\right)\right]} * \frac{\left(\frac{1}{\beta^{\alpha q}}\right)^n \Pi_{i=1}^n x_i^{\alpha q-1}\left[1+\left(\frac{x_i}{\beta}\right)^\alpha\right]^{-(q+1)} \alpha^{n+\rho_2-1}}{\int_0^\infty \left(\frac{1}{\beta^{\alpha q}}\right)^n \Pi_{i=1}^n x_i^{\alpha q-1}\left[1+\left(\frac{x_i}{\beta}\right)^\alpha\right]^{-(q+1)} \alpha^{n+\rho_2-1}d\alpha}.$$

$$\hat{B}_{EM-\alpha}(u) = \left[\int_0^\infty \frac{1}{\frac{1}{q}B_w\left[\left(q+\frac{1}{\alpha}\right),\left(1-\frac{1}{\alpha}\right)\right]} * \frac{\left(\frac{1}{\beta^{\alpha q}}\right)^n \Pi_{i=1}^n x_i^{\alpha q-1}\left[1+\left(\frac{x_i}{\beta}\right)^\alpha\right]^{-(q+1)} \alpha^{n+\rho_2-1}}{\int_0^\infty \left(\frac{1}{\beta^{\alpha q}}\right)^n \Pi_{i=1}^n x_i^{\alpha q-1}\left[1+\left(\frac{x_i}{\beta}\right)^\alpha\right]^{-(q+1)} \alpha^{n+\rho_2-1}d\alpha}\right]^{-1},$$

$$0 < u < 1.$$

$$(4.17)$$

Case III: Both the shape parameters α and q are unknown and β is known
Assuming both the shape parameters α and q to be independent, the joint distribution under Mukherjee-Islam prior is defined as

$$\pi_M^*(\alpha, q) = \frac{L(x|\alpha,q)*\pi_M(\alpha)*\pi_M(q)}{\iint_0^\infty L(x|\alpha,q)*\pi_M(\alpha)*\pi_M(q)d\alpha\ dq}$$

$$= \frac{\left(\frac{1}{\beta^{\alpha q}}\right)^n \Pi_{i=1}^n x_i^{\alpha q-1}\Pi_{i=1}^n\left[1+\left(\frac{x_i}{\beta}\right)^\alpha\right]^{-(q+1)} *\alpha^{n+\rho_2-1}q^{n+\rho_1-1}}{\iint_0^\infty \left(\frac{1}{\beta^{\alpha q}}\right)^n \Pi_{i=1}^n x_i^{\alpha q-1}\Pi_{i=1}^n\left[1+\left(\frac{x_i}{\beta}\right)^\alpha\right]^{-(q+1)} *\alpha^{n+\rho_2-1}q^{n+\rho_1-1}d\alpha\ dq}. \quad (4.18)$$

The Bayesian estimator of point measure of the Bonferroni curve $B_D(u)$, $0 < u < 1$, in the case of **Mukherjee-Islam prior** under **SELF**, is derived as

$$\hat{B}_{SM-\alpha q}(u) = \int_0^\infty B_D(u)*\pi_M^*(\alpha,q)d\alpha\ dq, \ 0 < u < 1$$

$$= \frac{\iint_0^\infty \frac{1}{q}B_w\left[\left(q+\frac{1}{\alpha}\right),\left(1-\frac{1}{\alpha}\right)\right] * \left(\frac{1}{\beta^{\alpha q}}\right)^n \Pi_{i=1}^n x_i^{\alpha q-1}\Pi_{i=1}^n\left[1+\left(\frac{x_i}{\beta}\right)^\alpha\right]^{-(q+1)} *\alpha^{n+\rho_2-1}q^{n+\rho_1-1}d\alpha\ dq}{\iint_0^\infty \left(\frac{1}{\beta^{\alpha q}}\right)^n \Pi_{i=1}^n x_i^{\alpha q-1}\Pi_{i=1}^n\left[1+\left(\frac{x_i}{\beta}\right)^\alpha\right]^{-(q+1)} *\alpha^{n+\rho_2-1}q^{n+\rho_1-1}d\alpha\ dq} \quad (4.19)$$

Similarly, for this case under the **ELF**, the Bayes estimator for the point measure $B_D(u)$, $0 < u < 1$ using **Mukherjee-Islam prior** is given as

$$\hat{B}_{EM\_\alpha q}(u) = [E\,(B^{-1}|x)]^{-1},\ 0 < u < 1$$

$$= \left[\frac{\iint_0^\infty \frac{1}{q} B_w\left[\left(q + \frac{1}{\alpha}\right),\left(1 - \frac{1}{\alpha}\right)\right] * \left(\frac{1}{\beta^{\alpha q}}\right)^n \Pi_{i=1}^n x_i^{\alpha q - 1} \Pi_{i=1}^n \left[1 + \left(\frac{x_i}{\beta}\right)^\alpha\right]^{-(q+1)} * \alpha^{n+\rho_2 - 1} q^{n+\rho_1 - 1} d\alpha\ dq}{\iint_0^\infty \left(\frac{1}{\beta^{\alpha q}}\right)^n \Pi_{i=1}^n x_i^{\alpha q - 1} \Pi_{i=1}^n \left[1 + \left(\frac{x_i}{\beta}\right)^\alpha\right]^{-(q+1)} * \alpha^{n+\rho_2 - 1} q^{n+\rho_1 - 1} d\alpha\ dq}\right]^{-1},$$

$$0 < u < 1,$$

$$(4.20)$$

where

$$E\,(B^{-1}|x) = \int_0^\infty \frac{1}{B_D(u)} * \pi_M^*(\alpha, q) d\alpha\ dq$$

$$= \frac{\iint_0^\infty \frac{1}{q} B_w\left[\left(q + \frac{1}{\alpha}\right),\left(1 - \frac{1}{\alpha}\right)\right] * \left(\frac{1}{\beta^{\alpha q}}\right)^n \Pi_{i=1}^n x_i^{\alpha q - 1} \Pi_{i=1}^n \left[1 + \left(\frac{x_i}{\beta}\right)^\alpha\right]^{-(q+1)} * \alpha^{n+\rho_2 - 1} q^{n+\rho_1 - 1} d\alpha\ dq}{\iint_0^\infty \left(\frac{1}{\beta^{\alpha q}}\right)^n \Pi_{i=1}^n x_i^{\alpha q - 1} \Pi_{i=1}^n \left[1 + \left(\frac{x_i}{\beta}\right)^\alpha\right]^{-(q+1)} * \alpha^{n+\rho_2 - 1} q^{n+\rho_1 - 1} d\alpha\ dq}.$$

Remark: The expressions cannot be simplified further, so the Bayesian estimators have been obtained using simulation techniques in R software as reported in Section 4.4.

4.3.2 BAYESIAN ESTIMATION OF POINT MEASURE OF BONFERRONI CURVE $B_D(u)$, $0 < u < 1$ UNDER UNIFORM PRIOR

Case I: Shape parameter q is unknown and α, β are known

The prior density of q using uniform prior is given as

$$\pi_U(q) \propto 1,\ 0 < q < \infty.$$

The posterior distribution of parameter q under uniform prior is

$$\pi_U^*(q|x) = \frac{T^{n+1} q^n e^{-qT}}{\Gamma(n+1)} \qquad (4.21)$$

The Bayes estimator for the point measure of $B_D(u)$, $0 < u < 1$, under **SELF**, in the case of **uniform prior** is derived as

$$\hat{B}_{SU\_q}(u) = \int_0^\infty B_D(u) * \pi_U^*(q|x) dq,\ 0 < u < 1,$$

$$= \frac{T^{n+1}}{\Gamma(n+1)} \int_0^\infty B_w\left[\left(q + \frac{1}{\alpha}\right),\left(1 - \frac{1}{\alpha}\right)\right] * q^{n-1} e^{-qT}\ dq. \qquad (4.22)$$

Similarly, under the **ELF,** the Bayes estimator of point measure $B_D(u)$, $0 < u < 1$, using **uniform prior** is given as

$$\hat{B}_{EU\_q}(u) = [E(B^{-1}|x)]^{-1}, \ 0 < u < 1,$$

$$[E(B^{-1}|x)]^{-1} = \left[\frac{T^{n+1}}{\Gamma(n+1)} \int_0^\infty B_w \left[\left(q + \frac{1}{\alpha} \right), \left(1 - \frac{1}{\alpha} \right) \right]^{-1} * q^{n+1} e^{-qT} dq \right]^{-1}.$$

$$\hat{B}_{EU\_q}(u) = \left[\frac{T^{n+1}}{\Gamma(n+1)} \int_0^\infty B_w \left[\left(q + \frac{1}{\alpha} \right), \left(1 - \frac{1}{\alpha} \right) \right]^{-1} * q^{n+1} e^{-qT} dq \right]^{-1}, \ 0 < u < 1.$$

$$(4.23)$$

Case II: Shape parameter α is unknown and q, β are known
  The prior density of α in this case is

$$\pi_U(\alpha) \propto 1, \ \ 0 < \alpha < \infty.$$

The posterior distribution of α is obtained as

$$\pi_U^*(\alpha) = \frac{L * \pi_U(\alpha)}{\int_0^\infty L * \pi_U(\alpha) da}, \ 0 < \alpha < \infty,$$

$$= \frac{\left(\frac{1}{\beta^{\alpha q}} \right)^n \Pi_{i=1}^n x_i^{\alpha q - 1} \left[1 + \left(\frac{x_i}{\beta} \right)^\alpha \right]^{-(q+1)} \alpha^n}{\int_0^\infty \left(\frac{1}{\beta^{\alpha q}} \right)^n \Pi_{i=1}^n x_i^{\alpha q - 1} \left[1 + \left(\frac{x_i}{\beta} \right)^\alpha \right]^{-(q+1)} \alpha^n \, d\alpha} \tag{4.24}$$

The Bayes estimator for the point measure $B_D(u)$, $0 < u < 1$, under **SELF** and **uniform prior** is derived as

$$\hat{B}_{SU\_\alpha}(u) = \int_0^\infty B_D(u) * \pi_U^*(\alpha) d\alpha, \ 0 < u < 1,$$

$$= \int_0^\infty \frac{1}{q} B_w \left[\left(q + \frac{1}{\alpha} \right), \left(1 - \frac{1}{\alpha} \right) \right] * \frac{\left(\frac{1}{\beta^{\alpha q}} \right)^n \Pi_{i=1}^n x_i^{\alpha q - 1} \left[1 + \left(\frac{x_i}{\beta} \right)^\alpha \right]^{-(q+1)} \alpha^n d\alpha}{\int_0^\infty \left(\frac{1}{\beta^{\alpha q}} \right)^n \Pi_{i=1}^n x_i^{\alpha q - 1} \left[1 + \left(\frac{x_i}{\beta} \right)^\alpha \right]^{-(q+1)} \alpha^n \, d\alpha}. \tag{4.25}$$

Similarly, under the **ELF,** the Bayes estimator for point measure $B_D(u)$, $0 < u < 1$ using **uniform prior** is given as

$$\hat{B}_{EU\_\alpha}(u)$$

$$= \left[\int_0^\infty \frac{1}{\frac{1}{q} B_w \left[\left(q + \frac{1}{\alpha} \right), \left(1 - \frac{1}{\alpha} \right) \right]} * \frac{\left(\frac{1}{\beta^{\alpha q}} \right)^n \Pi_{i=1}^n x_i^{\alpha q - 1} \left[1 + \left(\frac{x_i}{\beta} \right)^\alpha \right]^{-(q+1)} \alpha^n d\alpha}{\int_0^\infty \left(\frac{1}{\beta^{\alpha q}} \right)^n \Pi_{i=1}^n x_i^{\alpha q - 1} \left[1 + \left(\frac{x_i}{\beta} \right)^\alpha \right]^{-(q+1)} \alpha^n \, d\alpha} \right]^{-1},$$

$$0 < u < 1. \tag{4.26}$$

Case III: Shape parameters α and q are unknown and β is known

Assuming both the shape parameters α and q to be independent and following the **uniform prior,** the joint distribution is defined as

$$\pi_U^*(\alpha, q) = \frac{L(x \mid \alpha, q) * \pi_U(\alpha) * \pi_U(q)}{\iint_0^\infty L(x \mid \alpha, q) * \pi_U(\alpha) * \pi_U(q) d\alpha \ dq}, \quad \alpha, q > 0,$$

$$\pi_U^*(\alpha, q) = \frac{\left(\frac{1}{\beta^{\alpha q}}\right)^n \Pi_{i=1}^n x_i^{\alpha q - 1} \Pi_{i=1}^n \left[1 + \left(\frac{x_i}{\beta}\right)^\alpha\right]^{-(q+1)} \alpha^n q^n}{\iint_0^\infty \left(\frac{1}{\beta^{\alpha q}}\right)^n \Pi_{i=1}^n x_i^{\alpha q - 1} \Pi_{i=1}^n \left[1 + \left(\frac{x_i}{\beta}\right)^\alpha\right]^{-(q+1)} \alpha^n q^n d\alpha \ dq}. \qquad (4.27)$$

The Bayes estimator for the point measure $B_D(u)$, $0 < u < 1$, under **SELF** using **uniform prior** is derived as

$$\hat{B}_{SU\_\alpha q}(u) = \iint_0^\infty B_D(u) * \pi_U^*(\alpha, q) d\alpha \ dq, \ 0 < u < 1,$$

$$\hat{B}_{SU\_\alpha q}(u) = \iint_0^\infty \frac{1}{q} B_w \left[\left(q + \frac{1}{\alpha}\right), \left(1 - \frac{1}{\alpha}\right)\right] * \frac{\left(\frac{1}{\beta^{\alpha q}}\right)^n \Pi_{i=1}^n x_i^{\alpha q - 1} \Pi_{i=1}^n \left[1 + \left(\frac{x_i}{\beta}\right)^\alpha\right]^{-(q+1)} \alpha^n q^n d\alpha \ dq}{\iint_0^\infty \left(\frac{1}{\beta^{\alpha q}}\right)^n \Pi_{i=1}^n x_i^{\alpha q - 1} \Pi_{i=1}^n \left[1 + \left(\frac{x_i}{\beta}\right)^\alpha\right]^{-(q+1)} \alpha^n q^n d\alpha \ dq}.$$
$$(4.28)$$

Similarly, under the **ELF,** the Bayes estimator for the point measure of the Bonferroni curve $B_D(u)$, $0 < u < 1$, using **uniform prior** is given as

$$\hat{B}_{EU\_\alpha q}(u) = [E(B^{-1}|x)]^{-1}, \ 0 < u < 1.$$

$$E(B^{-1}|x) = \iint_0^\infty \frac{1}{\frac{1}{q} B_w \left[\left(q + \frac{1}{\alpha}\right), \left(1 - \frac{1}{\alpha}\right)\right]} * \frac{\left(\frac{1}{\beta^{\alpha q}}\right)^n \Pi_{i=1}^n x_i^{\alpha q - 1} \Pi_{i=1}^n \left[1 + \left(\frac{x_i}{\beta}\right)^\alpha\right]^{-(q+1)} \alpha^n q^n d\alpha \ dq}{\iint_0^\infty \left(\frac{1}{\beta^{\alpha q}}\right)^n \Pi_{i=1}^n x_i^{\alpha q - 1} \Pi_{i=1}^n \left[1 + \left(\frac{x_i}{\beta}\right)^\alpha\right]^{-(q+1)} \alpha^n q^n d\alpha \ dq}$$

Therefore,

$$\hat{B}_{EU\_\alpha q}(u)$$

$$= \left[\iint_0^\infty \frac{1}{\frac{1}{q} B_w \left[\left(q + \frac{1}{\alpha}\right), \left(1 - \frac{1}{\alpha}\right)\right]} * \frac{\left(\frac{1}{\beta^{\alpha q}}\right)^n \Pi_{i=1}^n x_i^{\alpha q - 1} \Pi_{i=1}^n \left[1 + \left(\frac{x_i}{\beta}\right)^\alpha\right]^{-(q+1)} \alpha^n q^n d\alpha \ dq}{\iint_0^\infty \left(\frac{1}{\beta^{\alpha q}}\right)^n \Pi_{i=1}^n x_i^{\alpha q - 1} \Pi_{i=1}^n \left[1 + \left(\frac{x_i}{\beta}\right)^\alpha\right]^{-(q+1)} \alpha^n q^n d\alpha \ dq}\right]^{-1},$$

$$0 < u < 1. \qquad (4.29)$$

4.3.3 BAYESIAN ESTIMATION OF POINT MEASURE OF BONFERRONI CURVE B_D (u), $0 < u < 1$, UNDER QUASI PRIOR

Case I: Shape parameter q is unknown and α, β are known

The prior density for the shape parameter q of the Dagum distribution under **quasi prior** is

$$\pi_Q(q) = \frac{1}{q^{d_1}}, \, q, \, d_1 > 0.$$

The posterior distribution of q is given by

$$\pi_Q^*(q|x) = \frac{L * \pi_Q(q)}{\int_0^\infty L * \pi_Q(q) dq}$$

$$= \frac{T^{n-d_1+1} q^{n-d_1} e^{-qT}}{\Gamma(n - d_1 + 1)}. \tag{4.30}$$

The Bayes estimator of point measure $B_D(u)$, $0 < u < 1$, under **SELF** for the **quasi prior** is

$$\hat{B}_{SQ-q}(u) = \int_0^\infty \frac{1}{q} B_w \left[\left(q + \frac{1}{\alpha} \right), \left(1 - \frac{1}{\alpha} \right) \right] * \frac{T^{n-d_1+1} q^{n-d_1} e^{-qT}}{\Gamma(n - d_1 + 1)} dq$$

$$= \frac{T^{n-d_1+1}}{\Gamma(n - d_1 + 1)} \int_0^\infty B_w \left[\left(q + \frac{1}{\alpha} \right), \left(1 - \frac{1}{\alpha} \right) \right] * q^{n-d_1-1} e^{-qT} \, dq. \tag{4.31}$$

The Bayes estimator of point measure $B_D(u)$, $0 < u < 1$ using **quasi prior** under the assumptions of **ELF** is derived as

$$\hat{B}_{EQ-q}(u) = [E(B^{-1}|x)]^{-1}, \, 0 < u < 1,$$

$$[E(B^{-1}|x)]^{-1} = \left[\frac{T^{n-d_1+1}}{\Gamma(n - d_1 + 1)} \int_0^\infty B_w \left[\left(q + \frac{1}{\alpha} \right), \left(1 - \frac{1}{\alpha} \right) \right]^{-1} * q^{n-d_1+1} * e^{-qT} dq \right]^{-1}.$$

Therefore,

$$\hat{B}_{EQ-q}(u) = \left[\frac{T^{n-d_1+1}}{\Gamma(n - d_1 + 1)} \int_0^\infty B_w \left[\left(q + \frac{1}{\alpha} \right), \left(1 - \frac{1}{\alpha} \right) \right]^{-1} * q^{n-d_1+1} * e^{-qT} dq \right]^{-1}. \tag{4.32}$$

Case II: Shape parameter α is unknown and q, β are known

The prior density under quasi prior for the shape parameter α is given as

$$\pi_Q(\alpha) = \frac{1}{\alpha^{d_2}}, \, \alpha, \, d_2 > 0.$$

The posterior distribution of α is derived as

$$\pi_Q^*(\alpha) = \frac{L * \pi_Q(\alpha)}{\int_0^\infty L * \pi_Q(\alpha) da}, \, \alpha, \, d_2 > 0$$

$$= \frac{\left(\frac{1}{\beta^{\alpha q}} \right)^n \Pi_{i=1}^n x_i^{\alpha q-1} \left[1 + \left(\frac{x_i}{\beta} \right)^\alpha \right]^{-(q+1)} \alpha^{n-d_2}}{\int_0^\infty \left(\frac{1}{\beta^{\alpha q}} \right)^n \Pi_{i=1}^n x_i^{\alpha q-1} \left[1 + \left(\frac{x_i}{\beta} \right)^\alpha \right]^{-(q+1)} \alpha^{n-d_2} \, d\alpha}.$$

The Bayes estimator for the point measure $B_D(u)$, $0 < u < 1$, under **SELF** using **quasi prior** is derived as

$$\hat{B}_{SQ-\alpha}(u) = \int_0^\infty \frac{1}{q} B_w \left[\left(q + \frac{1}{\alpha} \right), \left(1 \right. \right.$$
$$\left. \left. - \frac{1}{\alpha} \right) \right] * \frac{\left(\frac{1}{\beta^{\alpha q}} \right)^n \prod_{i=1}^n x_i^{\alpha q - 1} \left[1 + \left(\frac{x_i}{\beta} \right)^\alpha \right]^{-(q+1)} \alpha^{n-d_2} \, d\alpha}{\int_0^\infty \left(\frac{1}{\beta^{\alpha q}} \right)^n \prod_{i=1}^n x_i^{\alpha q - 1} \left[1 + \left(\frac{x_i}{\beta} \right)^\alpha \right]^{-(q+1)} \alpha^{n-d_2} \, d\alpha}. \qquad (4.33)$$

Similarly, under the **ELF,** the Bayes estimator for $B_D(u)$, $0 < u < 1$, using **quasi prior** is given as

$$\hat{B}_{EQ-\alpha}(u)$$
$$= \left[\int_0^\infty \frac{1}{\frac{1}{q} B_w \left[\left(q + \frac{1}{\alpha} \right), \left(1 - \frac{1}{\alpha} \right) \right]} * \frac{\left(\frac{1}{\beta^{\alpha q}} \right)^n \prod_{i=1}^n x_i^{\alpha q - 1} \left[1 + \left(\frac{x_i}{\beta} \right)^\alpha \right]^{-(q+1)} \alpha^{n-d_2} \, d\alpha}{\int_0^\infty \left(\frac{1}{\beta^{\alpha q}} \right)^n \prod_{i=1}^n x_i^{\alpha q - 1} \left[1 + \left(\frac{x_i}{\beta} \right)^\alpha \right]^{-(q+1)} \alpha^{n-d_2} \, d\alpha} \right]^{-1}.$$
$$(4.34)$$

Case III: Shape parameters α and q are unknown and β is known
Assuming both the shape parameters α and q to be independent following the **quasi prior** with the joint distribution defined as

$$\pi_Q^*(\alpha, q) = \frac{\left(\frac{1}{\beta^{\alpha q}} \right)^n \prod_{i=1}^n x_i^{\alpha q - 1} \prod_{i=1}^n \left[1 + \left(\frac{x_i}{\beta} \right)^\alpha \right]^{-(q+1)} \alpha^{n-d_1} q^{n-d_2}}{\iint_0^\infty \left(\frac{1}{\beta^{\alpha q}} \right)^n \prod_{i=1}^n x_i^{\alpha q - 1} \prod_{i=1}^n \left[1 + \left(\frac{x_i}{\beta} \right)^\alpha \right]^{-(q+1)} \alpha^{n-d_1} q^{n-d_2} d\alpha \, dq}.$$

the Bayes estimator for the point measure $B_D(u)$, $0 < u < 1$ under **SELF** using **quasi prior** is derived as

$$\hat{B}_{SQ-\alpha q}(u) = \iint_0^\infty B_D(u) * \pi_Q^*(\alpha, q) d\alpha \, dq, \quad 0 < u < 1$$
$$\hat{B}_{SQ-\alpha q}(u) = \iint_0^\infty \frac{1}{q} B_w \left[\left(q + \frac{1}{\alpha} \right), \left(1 \right. \right.$$
$$\left. \left. - \frac{1}{\alpha} \right) \right] * \frac{\left(\frac{1}{\beta^{\alpha q}} \right)^n \prod_{i=1}^n x_i^{\alpha q - 1} \prod_{i=1}^n \left[1 + \left(\frac{x_i}{\beta} \right)^\alpha \right]^{-(q+1)} \alpha^{n-d_1} q^{n-d_2} d\alpha \, dq}{\iint_0^\infty \left(\frac{1}{\beta^{\alpha q}} \right)^n \prod_{i=1}^n x_i^{\alpha q - 1} \prod_{i=1}^n \left[1 + \left(\frac{x_i}{\beta} \right)^\alpha \right]^{-(q+1)} \alpha^{n-d_1} q^{n-d_2} d\alpha \, dq} \qquad (4.35)$$

Similarly, under the **ELF,** the Bayes estimator of the point measure $B_D(u)$, $0 < u < 1$ using **quasi prior** is given as

$$\hat{B}_{EQ-\alpha q}(u) = \left[\iint_0^\infty \frac{1}{\frac{1}{q}B_w\left[\left(q+\frac{1}{\alpha}\right),\left(1-\frac{1}{\alpha}\right)\right]^*} \right.$$

$$\left. \frac{\left(\frac{1}{\beta^{\alpha q}}\right)^n \Pi_{i=1}^n x_i^{\alpha q-1} \Pi_{i=1}^n \left[1+\left(\frac{x_i}{\beta}\right)^\alpha\right]^{-(q+1)} \alpha^{n-d_1} q^{n-d_2} d\alpha\ dq}{\iint_0^\infty \left(\frac{1}{\beta^{\alpha q}}\right)^n \Pi_{i=1}^n x_i^{\alpha q-1} \Pi_{i=1}^n \left[1+\left(\frac{x_i}{\beta}\right)^\alpha\right]^{-(q+1)} \alpha^{n-d_1} q^{n-d_2} d\alpha\ dq} \right]^{-1}$$

$0 < u < 1$ (4.36)

As stated earlier, these expressions cannot be simplified further, so the Bayesian estimators have been obtained using simulation techniques in R software.

4.4 SIMULATION STUDY

The simulation study has been carried out in R software to examine the performance of Bayes estimators and estimated loss for the point measure of $B_D(u)$, $0 < u < 1$, for known and unknown shape parameters a and q of the Dagum distribution under an informative prior—Mukherjee-Islam prior—and two non-informative priors—uniform prior and quasi prior. In order to assess the performance of the above-mentioned methods used in estimation, different configurations of sample sizes are taken up for computation, along with some specific choices of parameters and hyper-parameters.

The process has been replicated 1,000 times, and the results are presented in Tables 4.1, 4.2, 4.3. For this purpose we choose $n = 25,\ 50,\ 100$—small, medium and large sample sizes—along with suitable choices of the shape parameters α and q and fixed the scale parameter $\beta = 1.85$. The value of u is chosen as 0.5 in the case of the Bonferroni curve $B_D(u)$. For the Mukerjee-Islam prior, the value of the hyper-parameters were chosen as $\rho = 3;\ \sigma = 2$.

4.4.1 CONCLUSION

The following observations can be drawn from Tables 4.1–4.3.

1. The estimated loss decreases as sample size n increases.
2. From all three tables it can be observed that the estimated loss is less in the case of Mukherjee-Islam prior as compared to quasi and uniform prior. The result holds true for small, medium and large sample sizes and also for different configurations of parameters a and q.
3. The choice of loss function depends upon the real-life situation. When one wants to take care of under- or over-estimation, then ELF is a suitable choice, and this results in less estimated loss than using SELF. This is particularly true in the scenario of income inequality where under-estimation gets more serious.

TABLE 4.1

Estimated Loss and Bayesian Estimates (In Parenthesis) Of Point Measure of Bonferroni Curve $B_D(u)$, $0 < u < 1$, when the Shape Parameter $q = 2, 3, \alpha = 7.5$, and $\beta = 1.85$ Using Mukherjee-Islam Prior, Uniform Prior and Quasi Prior Under Squared Error Loss Function and Entropy Loss Function (q unknown but α and β known)

n	q	$EL_{SM-q}(u)$	$EL_{EM-q}(u)$	$EL_{SU-q}(u)$	$EL_{EU-q}(u)$	$EL_{SQ-q}(u)$	$EL_{EQ-q}(u)$
25	2	0.160955	0.005465615	0.18014258	0.06201305	0.19006718	0.07247098
		(2.045686)	(2.4141084)	(2.493378)	(2.0439486)	(2.4100747)	(2.425136)
	3	0.1687939	0.03933489	0.16521496	0.05117569	0.17728351	0.05592904
		(2.094909)	(2.9823285)	(2.604284)	(2.8803944)	(3.0644053)	(2.4575773)
50	2	0.08766192	0.03629769	0.09609237	0.03854234	0.11929869	0.03965132
		(2.114417)	(2.0720156)	(2.5242639)	(2.3737184)	(2.0013617)	(2.3862024)
	3	0.1300512	0.01993288	0.1442371	0.02149543	0.1540964	0.02416738
		(2.575385)	(3.0053058)	(2.6795306)	(2.4387947)	(2.5873731)	(2.4554007)
100	2	0.06741986	0.02702488	0.06784082	0.03443066	0.0858466	0.02785943
		(2.609484)	(2.4375845)	(2.5249991)	(2.3800098)	(2.5058471)	(2.4526641)
	3	0.09131806	0.0150009	0.09449461	0.01575035	0.13134908	0.01924974
		(3.03696)	(3.0640875)	(3.0636992)	(3.0305807)	(3.0398092)	(2.4616066)

TABLE 4.2

Estimated Loss and Bayesian Estimates (In Parenthesis) of Bonferroni Curve $B_D(u)$, $0 < u < 1$, when the Shape Parameter $a = 2.5$, 3, $q = 1.5$ and $\beta = 1.85$ Using Mukherjee-Islam Prior, Uniform Prior and Quasi Prior Under Squared Error Loss Function and Entropy Loss Function (a unknown, q and β Known)

n	α	$EL_{SM\_\alpha}(u)$	$EL_{EM\_\alpha}(u)$	$EL_{SU\_\alpha}(u)$	$EL_{EU\_\alpha}(u)$	$EL_{SQ\_\alpha}(u)$	$EL_{SQ\_\alpha}(u)$
25	2.5	0.006740092	0.002863272	0.006928249	0.00297042	0.0077279	0.003653257
		(2.1664862)	(2.1664862)	(2.1661516)	(2.9742982)	(2.130287)	(2.1805296)
	3	0.006952446	0.00251345	0.00672881	0.002756404	0.00725463	0.00319548
		(3.191874)	(3.2632433)	(3.1914884)	(2.9742982)	(3.2867248)	(3.3389259)
50	2.5	0.00447027	0.002065014	0.005402933	0.002065014	0.005421343	0.003393524
		(2.1381846)	(2.1381846)	(2.1388171)	(2.9381846)	(2.1629864)	(2.1167057)
	3	0.003937583	0.0200050253	0.004448142	0.020810506	0.004803083	0.002662386
		(3.1592567)	(3.1592567)	(3.1599856)	(2.8956329)	(3.3502158)	(3.2251419)
100	2.5	0.00407657	0.001235271	0.004987506	0.001235271	0.004801361	0.002591666
		(3.09919366)	(2.09919366)	(2.1020901)	(2.85919366)	(2.1666005)	(2.1768017)
	3	0.00399369	0.0003665824	0.004281071	0.001130085	0.008144122	0.001400063
		(3.1143199)	(3.1143199)	(3.117658)	(2.840432)	(3.2839486)	(3.2574949)

TABLE 4.3

Estimated Loss and Bayesian Estimates (In Parenthesis) of Bonferroni Curve $B_D(u)$, $0 < u < 1$, when the Shape Parameter $a = 4, 7.5, q = 2.5, 5$ and $\beta = 1.85$ using Mukherjee-Islam Prior, Uniform Prior and Quasi Prior Under Squared Error Loss Function and Entropy Loss Function (a and q unknown, β Known)

	q	a	$EL_{SM\_cq}(u)$	$EL_{EM\_cq}(u)$	$EL_{SU\_cq}(u)$	$EL_{EU\_cq}(u)$	$EL_{SQ\_cq}(u)$	$EL_{EQ\_cq}(u)$
25	2.5	4	0.006740092	0.000804753	0.00792849	0.00693111	0.00904274	0.0085974.
			(3.1664862)	(3.752675)	(3.840230)	(3.869054)	(4.097798)	(3.871337)
	5	7.5	0.006952446	0.000701582	0.0071908	0.00596644	0.0092033	0.00583887
			(6.591874)	(6.800149)	(6.895287)	(6.882480)	(7.092398)	(6.778047)
50	2.5	4	0.00447027	0.000802650	0.00595864	0.00498721	0.00760667	0.00515071
			(3.1381846)	(3.800149)	(3.871296)	(3.892043)	(4.07546214)	(3.887255)
	5	7.5	0.003937583	0.000626766	0.00593128	0.00399619	0.007249318	0.00136318
			(6.5592567)	(6.860386)	(6.884036)	(6.896380)	(7.07971482)	(6.8993)
100	2.5	4	0.00407657	0.0007935735	0.00499701	0.00299998	0.00703042	0.0043853
			(4.09919366)	(3.800149)	(3.37126)	(3.869310)	(4.0887255)	(3.911592)
	5	7.5	0.00399369	0.000621066	0.00427416	0.00299991	0.00628847	0.000727108
			(6.9143199)	(6.861066)	(6.8290401)	(6.857414)	(7.1788698)	(6.9080399)

4.5 BAYESIAN ESTIMATOR OF POINT MEASURE OF ZENGA CURVE $Z_D(u)$, $0 < u < 1$, UNDER DIFFERENT PRIORS USING DIFFERENT LOSS FUNCTIONS

$Z_D(u)$, $0 < u < 1$–>Just like the Bonferroni curve $B_D(u)$, the Zenga curve $Z_D(u)$ also depends upon the shape parameters q and a only. In this section, the derivation of the point measure of Zenga curve $Z_D(u)$, $0 < u < 1$ is done in a similar fashion using informative and non-informative priors under the above-cited SELF (symmetrical loss function) and ELF (asymmetrical loss function) as mentioned in Section 4.3 for the Bonferroni curve $B_D(u)$. The results obtained are mentioned below.

Case I: Shape parameter q is unknown and α, β are known
Bayesian estimator of point measure of Zenga curve $Z_D(u)$, $0 < u < 1$, under **SELF** and **ELF** are given below.

Under Mukherjee-Islam prior:

$$\hat{Z}_{SM-q}(u) = \frac{T^{n+\rho_1}}{\Gamma(n+\rho_1)} \int_0^\infty \frac{q - B_w\left[\left(q+\frac{1}{\alpha}\right).\left(1-\frac{1}{\alpha}\right)\right]}{\left[1 - B_w\left[\left(q+\frac{1}{\alpha}\right).\left(1-\frac{1}{\alpha}\right)\right]\right]} q^{n+\rho_1-2} e^{-qT} \, dq, \, 0 < u < 1.$$

$$\hat{Z}_{EM-q}(u) = \left[\frac{T^{n+\rho_1}}{\Gamma(n+\rho_1)} \int_0^\infty \left[\frac{q - B_w\left[\left(q+\frac{1}{\alpha}\right).\left(1-\frac{1}{\alpha}\right)\right]}{\left[1 - B_w\left[\left(q+\frac{1}{\alpha}\right).\left(1-\frac{1}{\alpha}\right)\right]\right]}\right]^{-1} q^{n+\rho_1} {}_* e^{-qT} dq\right]^{-1}, \, 0 < u < 1.$$

Under uniform prior:

$$\hat{Z}_{SU-q}(u) = \frac{T^{n+1}}{\Gamma(n+1)} \int_0^\infty \frac{q - B_w\left[\left(q+\frac{1}{\alpha}\right).\left(1-\frac{1}{\alpha}\right)\right]}{\left[1 - B_w\left[\left(q+\frac{1}{\alpha}\right).\left(1-\frac{1}{\alpha}\right)\right]\right]} q^{n-1} e^{-qT} \, dq, \quad 0 < u < 1.$$

$$\hat{Z}_{EU-q}(u) = \left[\frac{T^{n+1}}{\Gamma(n+1)} \int_0^\infty \left[\frac{q - B_w\left[\left(q+\frac{1}{\alpha}\right).\left(1-\frac{1}{\alpha}\right)\right]}{\left[1 - B_w\left[\left(q+\frac{1}{\alpha}\right).\left(1-\frac{1}{\alpha}\right)\right]\right]}\right]^{-1} q^{n+1} {}_* e^{-qT} dq\right]^{-1}, \quad 0 < u < 1.$$

Under quasi prior:

$$\hat{Z}_{SQ-q}(u) = \frac{T^{n-d_1+1}}{\Gamma(n-d_1+1)} \int_0^\infty \frac{q - B_w\left[\left(q+\frac{1}{\alpha}\right).\left(1-\frac{1}{\alpha}\right)\right]}{\left[1 - B_w\left[\left(q+\frac{1}{\alpha}\right).\left(1-\frac{1}{\alpha}\right)\right]\right]} {}_* q^{n-d_1-1} e^{-qT} dq, \, 0 < u < 1.$$

$$\hat{Z}_{EQ-q}(u) = \left[\frac{T^{n-d_1+1}}{\Gamma(n-d_1+1)} \int_0^\infty \left[\frac{q - B_w\left[\left(q+\frac{1}{\alpha}\right).\left(1-\frac{1}{\alpha}\right)\right]}{\left[1 - B_w\left[\left(q+\frac{1}{\alpha}\right).\left(1-\frac{1}{\alpha}\right)\right]\right]}\right]^{-1} {}_* q^{n-d_1+1} e^{-qT} dq\right]^{-1}, \, 0 < u < 1.$$

Case II: Shape parameter α is unknown and q, β are known

Bayesian estimator of point measure of Zenga curve $Z_D(u)$, $0 < u < 1$, under **SELF** and **ELF** is given below.

Under Mukherjee-Islam prior:

$$\hat{Z}_{SM-\alpha}(u) = \int_0^\infty \frac{q - B_w\left[\left(q + \frac{1}{\alpha}\right),\left(1 - \frac{1}{\alpha}\right)\right]}{q\left[1 - B_w\left[\left(q + \frac{1}{\alpha}\right),\left(1 - \frac{1}{\alpha}\right)\right]\right]} * \frac{\left(\frac{1}{\beta^{\alpha q}}\right)^n \Pi_{i=1}^n x_i^{\alpha q - 1}\left[1 + \left(\frac{x_i}{\beta}\right)^\alpha\right]^{-(q+1)} \alpha^{n + \rho_2 - 1} d\alpha}{\int_0^\infty \left(\frac{1}{\beta^{\alpha q}}\right)^n \Pi_{i=1}^n x_i^{\alpha q - 1}\left[1 + \left(\frac{x_i}{\beta}\right)^\alpha\right]^{-(q+1)} \alpha^{n + \rho_2 - 1} d\alpha},$$

$0 < u < 1$.

$$\hat{Z}_{EM-\alpha}(u) = \left[\int_0^\infty \frac{1}{\frac{q - B_w\left[\left(q + \frac{1}{\alpha}\right),\left(1 - \frac{1}{\alpha}\right)\right]}{q\left[1 - B_w\left[\left(q + \frac{1}{\alpha}\right),\left(1 - \frac{1}{\alpha}\right)\right]\right]}} * \frac{\left(\frac{1}{\beta^{\alpha q}}\right)^n \Pi_{i=1}^n x_i^{\alpha q - 1}\left[1 + \left(\frac{x_i}{\beta}\right)^\alpha\right]^{-(q+1)} \alpha^{n + \rho_2 - 1} d\alpha}{\int_0^\infty \left(\frac{1}{\beta^{\alpha q}}\right)^n \Pi_{i=1}^n x_i^{\alpha q - 1}\left[1 + \left(\frac{x_i}{\beta}\right)^\alpha\right]^{-(q+1)} \alpha^{n + \rho_2 - 1} d\alpha}\right]^{-1},$$

$0 < u < 1$.

Under uniform prior:

$$\hat{Z}_{SU-\alpha}(u) = \int_0^\infty \frac{q - B_w\left[\left(q + \frac{1}{\alpha}\right),\left(1 - \frac{1}{\alpha}\right)\right]}{q\left[1 - B_w\left[\left(q + \frac{1}{\alpha}\right),\left(1 - \frac{1}{\alpha}\right)\right]\right]} * \frac{\left(\frac{1}{\beta^{\alpha q}}\right)^n \Pi_{i=1}^n x_i^{\alpha q - 1}\left[1 + \left(\frac{x_i}{\beta}\right)^\alpha\right]^{-(q+1)} \alpha^n d\alpha}{\int_0^\infty \left(\frac{1}{\beta^{\alpha q}}\right)^n \Pi_{i=1}^n x_i^{\alpha q - 1}\left[1 + \left(\frac{x_i}{\beta}\right)^\alpha\right]^{-(q+1)} \alpha^n d\alpha}, \quad 0 < u < 1.$$

$$\hat{Z}_{EU-\alpha}(u) = \left[\int_0^\infty \frac{1}{\frac{q - B_w\left[\left(q + \frac{1}{\alpha}\right),\left(1 - \frac{1}{\alpha}\right)\right]}{q\left[1 - B_w\left[\left(q + \frac{1}{\alpha}\right),\left(1 - \frac{1}{\alpha}\right)\right]\right]}} * \frac{\left(\frac{1}{\beta^{\alpha q}}\right)^n \Pi_{i=1}^n x_i^{\alpha q - 1}\left[1 + \left(\frac{x_i}{\beta}\right)^\alpha\right]^{-(q+1)} \alpha^n d\alpha}{\int_0^\infty \left(\frac{1}{\beta^{\alpha q}}\right)^n \Pi_{i=1}^n x_i^{\alpha q - 1}\left[1 + \left(\frac{x_i}{\beta}\right)^\alpha\right]^{-(q+1)} \alpha^n d\alpha}\right]^{-1}, \quad 0 < u < 1.$$

Under quasi prior:

$$\hat{Z}_{SQ-\alpha}(u) = \int_0^\infty \frac{q - B_w\left[\left(q + \frac{1}{\alpha}\right),\left(1 - \frac{1}{\alpha}\right)\right]}{q\left[1 - B_w\left[\left(q + \frac{1}{\alpha}\right),\left(1 - \frac{1}{\alpha}\right)\right]\right]} * \frac{\left(\frac{1}{\beta^{\alpha q}}\right)^n \Pi_{i=1}^n x_i^{\alpha q - 1}\left[1 + \left(\frac{x_i}{\beta}\right)^\alpha\right]^{-(q+1)} \alpha^{n - d_2} d\alpha}{\int_0^\infty \left(\frac{1}{\beta^{\alpha q}}\right)^n \Pi_{i=1}^n x_i^{\alpha q - 1}\left[1 + \left(\frac{x_i}{\beta}\right)^\alpha\right]^{-(q+1)} \alpha^{n - d_2} d\alpha}, \quad 0 < u < 1.$$

$$\hat{Z}_{EQ-\alpha}(u) = \left[\int_0^\infty \frac{1}{\frac{q - B_w\left[\left(q + \frac{1}{\alpha}\right),\left(1 - \frac{1}{\alpha}\right)\right]}{q\left[1 - B_w\left[\left(q + \frac{1}{\alpha}\right),\left(1 - \frac{1}{\alpha}\right)\right]\right]}} * \frac{\left(\frac{1}{\beta^{\alpha q}}\right)^n \Pi_{i=1}^n x_i^{\alpha q - 1}\left[1 + \left(\frac{x_i}{\beta}\right)^\alpha\right]^{-(q+1)} \alpha^{n - d_2} d\alpha}{\int_0^\infty \left(\frac{1}{\beta^{\alpha q}}\right)^n \Pi_{i=1}^n x_i^{\alpha q - 1}\left[1 + \left(\frac{x_i}{\beta}\right)^\alpha\right]^{-(q+1)} \alpha^{n - d_2} d\alpha}\right]^{-1}, \quad 0 < u < 1.$$

Case III: Shape parameters α and q are unknown and β is known

Bayesian estimator of point measure of Zenga curve $Z_D(u)$, $0 < u < 1$, under **SELF** and **ELF** is given below.

Under Mukherjee-Islam prior:

$$\hat{Z}_{SM\_\alpha q}(u)$$

$$= \frac{q - B_w\left[\left(q + \frac{1}{\alpha}\right), \left(1 - \frac{1}{\alpha}\right)\right] \left(\frac{1}{\beta^{\alpha q}}\right)^n \Pi_{i=1}^n x_i^{\alpha q - 1} \Pi_{i=1}^n \left[1 + \left(\frac{x_i}{\beta}\right)^\alpha\right]^{-(q+1)} * \alpha^{n+p_2-1} q^{n+p_1-1} d\alpha \; dq}{q\left[1 - B_w\left[\left(q + \frac{1}{\alpha}\right), \left(1 - \frac{1}{\alpha}\right)\right]\right] \iint_0^\infty \left(\frac{1}{\beta^{\alpha q}}\right)^n \Pi_{i=1}^n x_i^{\alpha q - 1} \Pi_{i=1}^n \left[1 + \left(\frac{x_i}{\beta}\right)^\alpha\right]^{-(q+1)} * \alpha^{n+p_2-1} q^{n+p_1-1} d\alpha \; dq},$$

$0 < u < 1$.

$$\hat{Z}_{EM\_\alpha q}(u)$$

$$= \left[\int_0^\infty \frac{1}{\frac{q - B_w\left[\left(q + \frac{1}{\alpha}\right), \left(1 - \frac{1}{\alpha}\right)\right]}{q\left[1 - B_w\left[\left(q + \frac{1}{\alpha}\right), \left(1 - \frac{1}{\alpha}\right)\right]\right]}} * \frac{\left(\frac{1}{\beta^{\alpha q}}\right)^n \Pi_{i=1}^n x_i^{\alpha q - 1} \Pi_{i=1}^n \left[1 + \left(\frac{x_i}{\beta}\right)^\alpha\right]^{-(q+1)} * \alpha^{n+p_2-1} q^{n+p_1-1} d\alpha \; dq}{\iint_0^\infty \left(\frac{1}{\beta^{\alpha q}}\right)^n \Pi_{i=1}^n x_i^{\alpha q - 1} \Pi_{i=1}^n \left[1 + \left(\frac{x_i}{\beta}\right)^\alpha\right]^{-(q+1)} * \alpha^{n+p_2-1} q^{n+p_1-1} d\alpha \; dq}\right]^{-1},$$

$0 < u < 1$.

Under uniform prior:

$$\hat{Z}_{SU\_\alpha q}(u) = \iint_0^\infty \frac{q - B_w\left[\left(q + \frac{1}{\alpha}\right), \left(1 - \frac{1}{\alpha}\right)\right]}{q\left[1 - B_w\left[\left(q + \frac{1}{\alpha}\right), \left(1 - \frac{1}{\alpha}\right)\right]\right]} * \frac{\left(\frac{1}{\beta^{\alpha q}}\right)^n \Pi_{i=1}^n x_i^{\alpha q - 1} \Pi_{i=1}^n \left[1 + \left(\frac{x_i}{\beta}\right)^\alpha\right]^{-(q+1)} \alpha^n q^n d\alpha \; dq}{\iint_0^\infty \left(\frac{1}{\beta^{\alpha q}}\right)^n \Pi_{i=1}^n x_i^{\alpha q - 1} \Pi_{i=1}^n \left[1 + \left(\frac{x_i}{\beta}\right)^\alpha\right]^{-(q+1)} \alpha^n q^n d\alpha \; dq},$$

$0 < u < 1$

$$\hat{Z}_{EU\_\alpha q}(u) = \left[\iint_0^\infty \frac{1}{\frac{q - B_w\left[\left(q + \frac{1}{\alpha}\right), \left(1 - \frac{1}{\alpha}\right)\right]}{q\left[1 - B_w\left[\left(q + \frac{1}{\alpha}\right), \left(1 - \frac{1}{\alpha}\right)\right]\right]}} * \frac{\left(\frac{1}{\beta^{\alpha q}}\right)^n \Pi_{i=1}^n x_i^{\alpha q - 1} \Pi_{i=1}^n \left[1 + \left(\frac{x_i}{\beta}\right)^\alpha\right]^{-(q+1)} \alpha^n q^n d\alpha \; dq}{\iint_0^\infty \left(\frac{1}{\beta^{\alpha q}}\right)^n \Pi_{i=1}^n x_i^{\alpha q - 1} \Pi_{i=1}^n \left[1 + \left(\frac{x_i}{\beta}\right)^\alpha\right]^{-(q+1)} \alpha^n q^n d\alpha \; dq}\right]^{-1},$$

$0 < u < 1$

Under quasi prior:

$\hat{Z}_{SQ-\alpha q}(u)$

$$= \int\!\!\int_0^\infty \frac{q - B_w\left[\left(q + \frac{1}{\alpha}\right), \left(1 - \frac{1}{\alpha}\right)\right]}{q\left[1 - B_w\left[\left(q + \frac{1}{\alpha}\right), \left(1 - \frac{1}{\alpha}\right)\right]\right]} * \frac{\left(\frac{1}{\beta^{\alpha q}}\right)^n \Pi_{i=1}^n x_i^{\alpha q - 1} \Pi_{i=1}^n \left[1 + \left(\frac{x_i}{\beta}\right)^\alpha\right]^{-(q+1)} \alpha^{n - d_1} q^{n - d_2}\, d\alpha\ dq}{\int\!\!\int_0^\infty \left(\frac{1}{\beta^{\alpha q}}\right)^n \Pi_{i=1}^n x_i^{\alpha q - 1} \Pi_{i=1}^n \left[1 + \left(\frac{x_i}{\beta}\right)^\alpha\right]^{-(q+1)} \alpha^{n - d_1} q^{n - d_2} d\alpha\ dq},$$

$0 < u < 1$.

$\hat{Z}_{EQ-\alpha q}(u)$

$$= \left[\int\!\!\int_0^\infty \frac{1}{\dfrac{q - B_w\left[\left(q + \frac{1}{\alpha}\right), \left(1 - \frac{1}{\alpha}\right)\right]}{q\left[1 - B_w\left[\left(q + \frac{1}{\alpha}\right), \left(1 - \frac{1}{\alpha}\right)\right]\right]}} * \frac{\left(\frac{1}{\beta^{\alpha q}}\right)^n \Pi_{i=1}^n x_i^{\alpha q - 1} \Pi_{i=1}^n \left[1 + \left(\frac{x_i}{\beta}\right)^\alpha\right]^{-(q+1)} \alpha^{n - d_1} q^{n - d_2} d\alpha\ dq}{\int\!\!\int_0^\infty \left(\frac{1}{\beta^{\alpha q}}\right)^n \Pi_{i=1}^n x_i^{\alpha q - 1} \Pi_{i=1}^n \left[1 + \left(\frac{x_i}{\beta}\right)^\alpha\right]^{-(q+1)} \alpha^{n - d_1} q^{n - d_2} d\alpha\ dq} \right]^{-1},$$

$0 < u < 1$.

4.6   SIMULATION STUDY

Using the same set of configurations as mentioned in Section 4.4 for the sample size n and the shape parameters q and α of the Dagum distribution, the values of estimated loss and Bayesian estimator for point measure of the Zenga curve $Z_D(u)$, $0 < u < 1$, have been given in the tables below (Tables 4.4–4.6).

The following observations can be drawn from the Tables 4.4–4.6.

1. The estimated loss in each case decreases as sample size n increases.
2. The estimated loss corresponding to ELF is less compared to SELF under all three priors. Therefore, it can be concluded that ELF is superior to SELF for obtaining the Bayesian estimator of point measure of Zenga curve $Z_D(u)$, $0 < u < 1$, for given values of shape parameters α and q keeping scale parameter β fixed.
3. Also, from all three tables it can be observed that the estimated loss is less in Mukherjee-Islam prior as compared to quasi and uniform priors.

4.7   REAL-LIFE EXAMPLE

For illustration of the above proposed estimators, consider the real-life data depicting the unemployment rate for 50 countries (Trading economics 2019 https://tradingeconomics.com/country-list/unemployment-rate).

We use the easy fit software to show that the data set fits well to the Dagum distribution with p-value for the Kolmogorov-Smirnov test equal to 0.37651 at a 5% level of significance. The estimated loss of Bayes estimate of Bonferroni curve $B_D(u)$ and Zenga curve $Z_D(u)$ at specific values of u, $0 < u < 1$ are obtained for this data under two loss functions using three different priors. The values of

TABLE 4.4

Estimated Loss and Bayesian Estimates (In Parenthesis) of Point Measure of Zenga Curve $Z_D(u)$, $0 < u < 1$, when the Shape Parameter $q = 2$, 3, $a = 7.5$ and $\beta = 1.85$ Using Mukherjee-Islam Prior, Uniform Prior and Quasi Prior Under Squared Error Loss Function and Entropy Loss Function

n	q	$EL_{SM-q}(u)$	$EL_{EM-q}(u)$	$EL_{SU-q}(u)$	$EL_{EU-q}(u)$	$EL_{SQ-q}(u)$	$EL_{EQ-q}(u)$
25	2	0.01030438	0.0001099686	0.01056634	0.0001130735	0.7725006	0.00979589
		(1.9249832)	(2.090646)	(1.9232239)	(1.090985)	(2.064879)	(1.9312131)
	3	0.01388535	0.0002152898	0.01643371	0.0002349016	0.9734412	0.01530836
		(2.9291129)	(3.077614)	(2.8147898)	(1.082365)	(3.058043)	(2.921893)
50	2	0.01013763	6.060164e-05	0.01049741	8.606142e-05	0.515336	0.00940579
		(1.926116)	(2.088033)	(1.9236845)	(2.078141)	(2.066574)	(1.9220919)
	3	0.01439996	0.0001605156	0.01418825	0.000133666	0.8816978	0.01336925
		(2.9261084)	(3.064659)	(2.977337)	(3.073316)	(3.082828)	(2.8609485)
100	2	0.01089124	3.634218e-05	0.01098819	3.975363e-05	0.388159	0.00807368
		(1.9204405)	(2.080029)	(2.9210749)	(2.083097)	(2.089299)	(1.92847)
	3	0.01480142	6.048484e-05	0.01531976	6.854098e-05	0.8385787	0.01311945
		(2.9208848)	(3.071457)	(3.9238061)	(3.076864)	(3.065814)	(2.987021)

TABLE 4.5

Estimated Loss and Bayesian Estimates (In Parenthesis) of Zenga Curve $Z_D(u)$, $0 < u < 1$, when the Shape Parameter $a = 2.5$, 3, $q = 1.5$ and $\beta = 1.85$ Using Mukherjee-Islam Prior, Uniform Prior and Quasi Prior Under Squared Error Loss Function and Entropy Loss Function

n	a	$EI_{SM\_\alpha}(u)$	$EL_{EM\_\alpha}(u)$	$EI_{SU\_\alpha}(u)$	$EL_{EU\_\alpha}(u)$	$EI_{SQ\_\alpha}(u)$	$EI_{EQ\_\alpha}(u)$
25	2.5	0.02879123	0.002863272	0.0291263	0.003180235	0.03110816	0.003969799
		(2.188116)	(2.288116)	(2.140123)	(2.205845)	(2.08593)	(2.262581)
	3	0.03008547	0.004347321	0.0304356	0.002192024	0.03250656	0.006102992
		(3.002304)	(3.085667)	(3.065467)	(3.079623)	(3.010069)	(2.8972005)
50	2.5	0.1880773	0.002065014	0.0883309	0.002435551	0.1909537	0.003481933
		(2.303143)	(2.3031438)	(2.12871)	(2.3457323)	(2.1672775)	(2.2863374)
	3	0.1965319	0.002546093	0.09230161	0.003609194	0.1995376	0.005411749
		(3.009232)	(3.035544)	(3.051578)	(2.8130988)	(2.9274456)	(2.8670632)
100	2.5	0.3022854	0.00123527	0.2732009	0.001130365	0.2984809	0.01813553
		(2.348307)	(2.3483076)	(2.1505987)	(2.3650746)	(2.3672382)	(2.3764352)
	3	0.3158739	0.00015865	0.2854821	0.001465764	0.3229207	0.01811364
		(3.106271)	(3.058573)	(2.767284)	(2.639412)	(2.6940176)	(2.7464755)

TABLE 4.6

Estimated Loss and Bayesian Estimate (In Parenthesis) of Zenga Curve $Z_D(u)$, $0 < u < 1$ When the Shape Parameter $a = 4, 7.5$, $q = 2.5, 5$ and $\beta = 1.85$ Using Mukherjee-Islam Prior, Uniform Prior and Quasi Prior Under Squared Error Loss Function and Entropy Loss Function

n	q	a	$EL_{SM\_aq}(u)$	$EL_{EM\_aq}(u)$	$EL_{SU\_aq}(u)$	$EL_{EU\_aq}(u)$	$EL_{SQ\_aq}(u)$	$EL_{EQ\_aq}(u)$
25	2.5	4	0.1070431	0.035848	0.1787499	0.339551	0.2904799	0.39848
			(3.08328391)	(3.05607513)	(3.08147181)	(3.04794603)	(3.05389618)	(3.05607513)
	5	7.5	0.1184155	0.0394814	0.19956483	0.416314	0.3643088	0.417741
			(7.08885024)	(7.04756776)	(7.07245684)	(7.04607579)	(7.060358)	(7.05391141)
50	2.5	4	0.1048117	0.0104741	0.17278983	0.114741	0.1828636	0.197231
			(3.08477879)	(3.05391141)	(3.06966101)	(3.05391141)	(3.03359518)	(3.05734907)
	5	7.5	0.07494266	0.0163432	0.1998117	0.363432	0.1998895	0.3638532.
			(7.07068364)	(7.05255396)	(7.08477879)	(7.05255396)	(7.03726872)	(7.05617747)
100	2.5	4	0.07278983	0.008010963	0.1612037	0.05163	0.1700238	0.169603
			(3.06966101)	(3.1123886)	(3.03475665)	(3.09656817)	(3.1303932)	(3.06733775)
	5	7.5	0.03470518	0.005437	0.17470518	0.8010963	0.16362803	0.3443712.
			(7.04810072)	(7.08120981)	(7.04810072)	(7.1123886)	(7.1537141)	(7.08120981)

TABLE 4.7
Estimated Loss for Bonferroni Curve $B_D(u)$, $0 < u < 1$, Under Real-Life Data with $q = 2.43$, $a = 4.226$, $\beta = 2.15$, $n = 50$ and $u = 0.5$

Loss Functions	Mukherjee-Islam Prior	Uniform Prior	Quasi Prior
Squared Error Loss Function	0.1889579	0.2127188	0.327295
Entropy Loss Function	0.1497411	0.4115226	0.5497411

TABLE 4.8
Estimated Loss for Zenga Curve $Z_D(u)$, $0 < u < 1$ Under Real-Life Data

Loss Functions	Mukherjee-Islam Prior	Uniform Prior	Quasi Prior
Squared Error Loss Function	0.5208778	0.5529809	0.656666
Entropy Loss Function	0.2220625	0.2903656	0.4035862

estimated loss are given below in Tables 4.7 and 4.8, where sample size $n = 50$, $u = 0.50$, $q = 2.43$, $\alpha = 4.226$, $\beta = 2.15$.

As can be observed from Tables 4.7 and 4.8, the results obtained from the simulation of the real-life example are in accordance with those of the simulation study. The estimates using Mukherjee-Islam prior seem to be the best in the case of the real data set as well because these result in the smallest estimated loss for both Bonferroni and Zenga curves.

REFERENCES

Aaberge, R. (2000). Characterizations of Lorenz curves and income distributions. *Social Choice and Welfare, 17*, 639–653.

Ahmed, A., Naqash, S., & Ahmad, S. P. (2017). Bayesian analysis of Dagum distribution. *Journal of Reliability and Statistical Studies,* ISSN (Print), 0974-8024, (Online): 2229-5666, *10*(1), 123–136.

Arora, S., Mahajan, K. K., & Godura, P. (2014, December). Bayesian estimation of Zenga curve using different priors in case of pareto distribution, *International Journal of Statistics - Theory and Applications, 1*(1), 1–7.

Atkinson, A. B. (1970). On the measurement of inequality. *Journal of Economic Theory, 2*, 244–263.

Bonferroni, C. E. (1930). *Elementi di Statistca generale.* Seeber.

Burr, I. W. (1942). Cumulative frequency functions. *Annals of Mathematical Statistics, 13*(2), 215–232. http://dx.doi.org/10.1214/aoms/1177731607

Calabria,R., & Pulcini,G. (1990). On the maximum likelihood and least-squares estimation in the Inverse Weibull distributions. *Statistical Application, 2*(1), 53–66.

Chotikapanich, D. Griffiths, W. E. (2006). Bayesian assessment of Lorenz and Stochastic dominance in income distributions. *Research on Economic Inequality, 13*, 297–321.

Dagum, C. (1975). A model of income distribution and the conditions of existence of moments of finite order, Bulletin of the International Statistical Institute. In *46 Proceedings of the 40th Session of the ISI, Warsaw, Contributed Papers* (pp. 199–205).

Dagum, C. (1983). Income distribution models. In S. Kotz, N. L. Johnson, & C. Read (Eds.), *Encyclopedia of statistical sciences* (Vol. 4, pp. 27–34). John Wiley.

Dey, D. K., Ghosh, M. & Srinivasan, C. (1987). Simultaneous estimation of parameters under entropy loss. *Journal of Statistical Planning Inference, 15*, 347–363.

Domma, F. (2007). Asymptotic distribution of the maximum likelihood estimators of the parameters of the right-truncated Dagum distribution. *Communications in Statistics-Simulation and Computation, 36*(6), 1187–1199.

Domma, F., Giordano. S., & Zenga, M. M. (2011). Maximum likelihood estimation in Dagum distribution with censored sample. *Journal of Applied Statistics, 38*(12), 2971–2985.

Garvey, R. G., Castillo, E., Cavalier, T. M., & Lehtihet, E. (2002). Four-parameter beta distribution estimation and skewness test. *Quality and Reliability Engineering International, 18*(5), 395–402.

Giorgi, G. M., & Crescenzi, M. (2001). A look at the Bonferroni inequality measure in a reliability framework. *Statistica, 61*(4), 571–583.

Klefsjö, B. (1984). Reliability interpretations of some concepts from economics. *Naval Research Logistics Quarterly, 31*,301–308.

Kleiber C., & Kotz S. (2003). *Statistical size distributions in economics and actuarial sciences.* Wiley.

Lorenz, M. O. (1905). Methods of measuring the concentration of wealth. *Journal of the American Statistical Association (New Series), 70*, 209–217.

Mahmoud, A. W., Ghazal, M. G. M., & Radwan, H. M. M.(2017). Inverted generalized linear exponential distribution as a lifetime model. *Applied Mathematics and Information Sciences, 11*(6), 1747–176.

Mukheerji, S. P. & Islam, A. (1983). A finite range distribution of failures times. *Naval Research Logistics Quarterly, 30*, 487–491.

Muliere,P., & Scarsini,M. (1989). A note on stochastic dominance and inequality measures. *Journal of Economic Theory, 49*, 314–323.

Sathar, E. I. A., Jeevanand, E. S.,& Nair, K. R. M. (2005). Bayesian estimation of Lorenz curve, Gini – Index and variance of logarithms in a pareto distribution. *Statistica anno, 65* (2), 193–205.

Zenga, M. (2007). Inequality curve and inequality index based on the ratios between lower and upper arithmetic means, statistica and applicazioni (5), 3-27.R. (2000a). *Characterizations of Lorenz Curves and Income Distributions, Social Choice and Welfare, 17*, 639–653.

5 Band Pass Filters and their Applications in Time Series Analyses

Bhasha H. Vachharajani[1], Dency V. Panicker[2], and Poonam Mishra[1]
[1]Department of Mathematics, School of Technology, Pandit Deendayal Petroleum University, Gandhinagar, Gujarat
[2]Department of Physics, School of Technology, Pandit Deendayal Petroleum University, Gandhinagar, Gujarat

5.1 INTRODUCTION: TIME SERIES

To develop a forecast model, we suggest beginning with a graphical display and analysis of the available data, as apparent features of the data are captured. Visual tools are most effective, though we do not neglect the usefulness of analytical tools. However, as per the statement by Yogi Berra, "You can observe a lot by just watching", observation is important prior to arriving at any conclusion. Time series data are quantities that are collected at various timesteps and arranged chronologically. The difference between two timesteps may be uniform or non-uniform. Analyzing a time series is best suitable for studying a dynamic process. Time series analyses cover almost all the fields, ranging from economics to finance, social science to physical science. Gross Domestic Product (GDP), Consumer Price Index (CPI), birth rate, mortality rate, population, blood pressure tracking, heart rate monitoring, global temperature, and pollution levels are a few examples that make use of time series data.

Time series data may be listed in terms of ordered pairs, just like a mathematical function, (t, f(t)), where t denotes time and f(t) is the value of the quantity at time t. Data may be represented in tabular form, with two columns. However, the simplest form to visualize and study time series data is to plot the data on a graph. The graph may be a line plot, a histogram or a scatter plot. However, it is interesting to note that some of the classical tools of descriptive statistics, such as the histogram and the stem-and-leaf display, are not particularly useful for time series data because they do not take time into account. There are many variations of the time series plot and other graphical displays that can be constructed to show specific features of a data set.

Just by observing the graph of data, one can make out whether it is regular or irregular. Further, if it is regular, does it follow a specific trend (rising or declining)

DOI: 10.1201/9781003102281-5

or possess a periodic nature? All such questions can be answered just by looking at the graph. Further, one can study variations (annual/seasonal) and relate parameters through a time series. For carrying out spatio-temporal analyses of a parameter, time series can be extremely useful. We introduce types of time series mentioned above and explain their corresponding graph one by one.

- **TREND:** Identifying trend in a time series involves finding long-term changes in its average behavior. Such trend analysis helps identify noticeable changes in the system that occur over a longer period of time. For example, sea ice extent in the Arctic is found to follow a DECREASING trend (Figure 5.1) for the last three decades, whereas in the Antarctic, there's an INCREASING trend. (Figure 5.1).
- **PERIODIC:** At times, we may be able to spot some particular patterns, which are reiterated after a specific time. Such series are said to follow a PERIODIC trend (Figure 5.2). Such patterns help identifying common or contrasting features occurring within a time series. In addition to this, if the patterns are found to repeat year after year, we say such graphs exhibit a SEASONAL trend. When the patterns (rise and fall) are found to repeat, not necessarily after a fixed tenure, we say such graphs have a CYCLIC nature.
- **IRREGULAR/FLUCTUATING:** If no specific trend or pattern is revealed and no conclusion regarding the nature of the time series can be drawn, the graph is said to be IRREGULAR (Figure 5.3).

For a fluctuating time series, the variations are random, and we regard the data as "noisy". Let us now study ways to get rid of this "noise". At times, the value of the quantity may fall out of the desired interval just for a finite amount of timesteps (Figure 5.4). Such a case arises when there is a failure of data measurement. We simply replace it by some "fixed value" falling in the range or interpolate with the available observations.

One of the major applications of time series analysis is predicting the future, generally termed *forecasting*, which is attained through modeling. The unusual fluctuations of the data set perturb this feature of time series, and hence the forecast becomes difficult. However, with some of the techniques, one can rule out these unwanted data points, thereby improving the forecast capability of the model.

5.1.1 SMOOTHING

Smoothing is one of the techniques of filtering (discussed in the next section). In order to see a clear signal or pattern in time series data and remove irregular roughness, we apply smoothing techniques. In the case of seasonal data, seasonality is made smooth in order to identify its trend. At times, there might be some ambiguity in the data, which is resolved by suitable mathematical techniques. For example, in the case of missing data, we apply interpolation using the surrounding points.

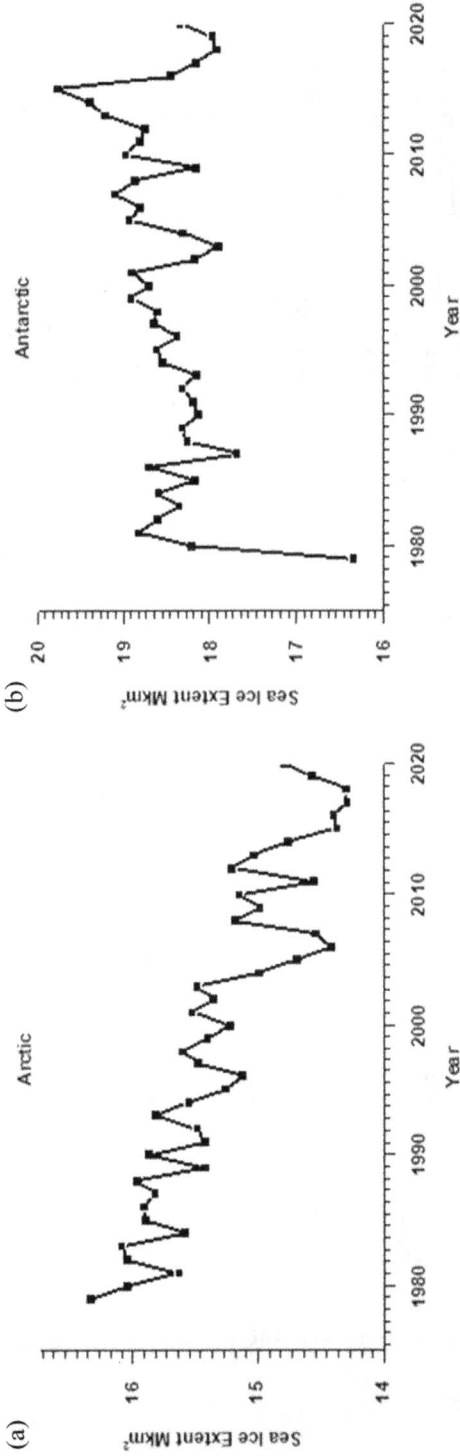

FIGURE 5.1 (a) Sea ice extent (million km²) over the Arctic for the month of March (maximum) (b) Sea ice extent over Antarctica for the month of September (maximum) for the span 1979–2020 (NSIDC).

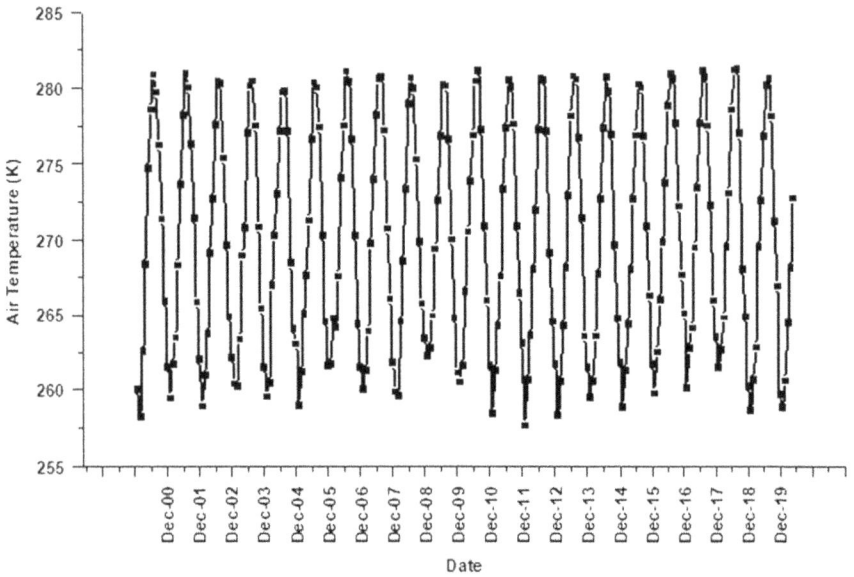

FIGURE 5.2 Air temperature (2 m) over the Himalayan region for the span 2000–2020 (ERAECMWF).

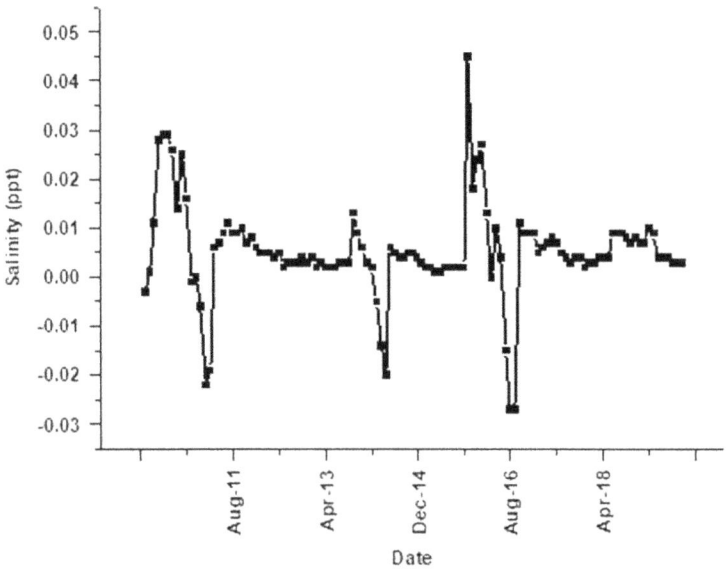

FIGURE 5.3 Salinity (ppt) over the East Siberian Sea for the span 2010–2019 (ERA5-ECMWF).

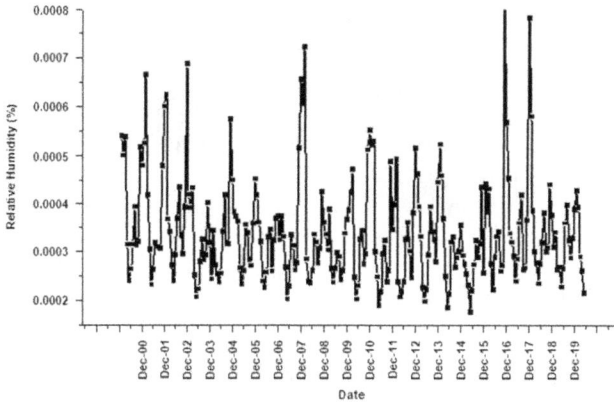

FIGURE 5.4 Relative humidity (%) over the Himalayan region for the span 2000–2020 (ERA5-ECMWF).

5.2 FILTERING

Most of the signals we work with include data that is slightly erroneous or contains some unwanted signals (also referred to as "noise"). For carrying out signal processing and analysis, it is imperative to get rid of such interferences or at least reduce their effects. To extract information from noisy data, we apply filtering techniques. The objective of filtering is to estimate the true state and/or predicting future value. A filter is a system which allows passing (or amplifying) certain frequencies while reducing (attenuating) other frequencies. It rules out (ignores) unwanted frequencies from signals and thus helps extract desired frequencies.

The four broad categories of filters we will study are low pass filter, high pass filter, band pass filter and band stop filter. We will discuss each type in detail.

5.2.1 LOW PASS FILTER

As the name suggests, a low pass filter allows low frequencies to pass through and attenuates greater frequencies. Its graph looks like the one depicted in Figure 5.5.

The graph indicates that low frequencies are allowed to pass through and high frequencies are ignored or reduced. Cut-off frequency is the one above which the frequencies are attenuated. Note that we have defined the low pass filter here by frequency and not wavelength. If we consider wavelength, then the description will be reversed, that is, a low pass filter will retain higher wavelengths and ignore lower wavelengths. Shirahata et al. (2016) have discussed the role of low pass filters to remove tidal effects from ground water observations data.

Low pass filtering of a time series reveals low-frequency features and trends.

Low pass filters are used in eliminating background noise, play a role in tuning radio to a specific frequency, modify digital images and remove specific frequencies in data analysis.

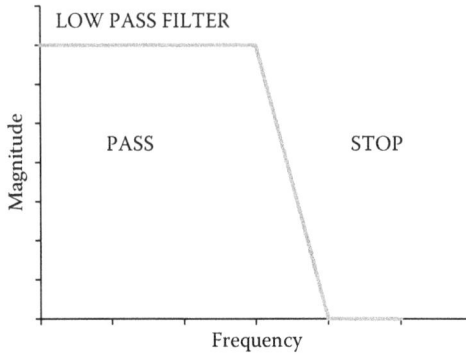

FIGURE 5.5 Low pass filter.

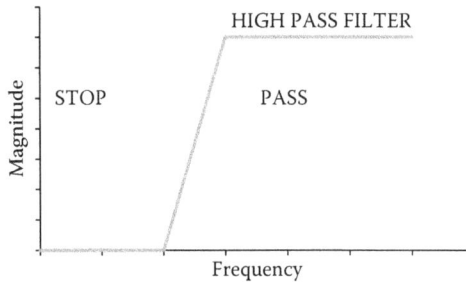

FIGURE 5.6 High pass filter.

5.2.2 HIGH PASS FILTER

This category of filters allows high frequencies to pass through and blocks low frequencies. Figure 5.6 represents a graph of a typical high pass filter.

Here, the cut-off frequency is the one below which the frequencies are attenuated.

High pass filtering of a time series isolates high frequency transients, for example, storms.

A few applications of high pass filters include loud speakers (noise at a low level is attenuated) and audio amplifiers (higher frequencies are amplified).

We now discuss some most commonly used filters (Montgomery et al., 2008;Peng, 2020), they have been formed using either low pass or high pass filters:

- **Moving average filter (MAF) (m):**

A moving average filter (MAF) is a simple low pass filter which computes the average of a fixed number of input signals at a time. It is a method of smoothing images by reducing the variation between neighboring pixels. In an image, we move pixel-wise, replace each pixel value by the average value of neighboring pixels, including the pixel itself. Let us understand this mathematically.

Suppose we have the observed values as $(x_1, \ x_2, \ x_3, \ ...., x_{m-1}, x_m, x_{m+1}, \ ....)$ at timesteps $(t_1, \ t_2, \ t_3, \ ...., t_{m-1}, t_m, t_{m+1}, \ ....)$. We modify this series a bit. Say we take the average of $x_1, x_2 \ and \ x_3$ as the first value (denote it by y_1); the average of $x_2, x_3 \ and \ x_4$ (denote it by y_2) as the second value and so on. This process is continued till the last term is exhausted. Putting this in words, we have considered the running mean of length 3 to the original series. This would filter out the distortions present in the original data and thus smooth the data. Let us generalize the above process. We form a new series by applying a running mean of length m to the original series, which is termed as the MAF of length m, i.e., (MAF) (m). Its mathematical form is:

$$y_t = \frac{1}{m} \Sigma_{k=-m/2}^{k=m/2} \ x_{t+k}, \ \text{if m is even,}$$

$$y_t = \frac{1}{m} \Sigma_{k=-(m-1)/2}^{k=(m-1)/2} \ x_{t+k}, \ \text{if m is odd, m} > 1; (x_t, y_t) \ \text{is the point, corresponding to}$$
timestep t.

For instance, the four-month running mean filter MAF (4) coarsely filters out the intraseasonal oscillations; hence, it is used to study seasonal patterns.

- **Binomial filters $(1/2,1/2)^k$:**

These filters are derived from MAF. Here, we repetitively apply the MAF (2) that has weights (1/2, 1/2). For example, with k = 3, these weights become $(1/2, 1/2)^3 =$ (1,3,3,1)/8, which reduce to zero smoothly as we move near the edges.

Note that the weights assigned to a binomial filter correspond to the coefficients of binomial expansion. Hence, we may choose a row from Pascal's triangle for generating the weights, i.e., for k = 4, we have to choose the $(k+1)^{th}$ row of Pascal's triangle. With an increase in k, the terms are likely to approach zero in a gradual manner.

- **Exponential smoothers:**

Let us understand the need of an exponential smoother first. For a time series with a clear pattern, MAF works. However, if the patterns are not revealed clearly, exponential filters have to be applied. MAF gives equal weight to past data, thereby giving an equal impact to these data points for future forecasting. In general, the recent past might contribute more to the present data than the longer past. Hence, we may need to assign more weight to the recent past than to the distant past. This concept of varying weights works behind the exponential smoothers.

The most basic exponential smoother is the single/simple exponential smoother. Here, the weights decrease exponentially as we move backward in time frame. The smallest weights are assigned to the oldest observations, and higher weights are attached to the most recent observations.

For forecast purposes, the exponential smoothing filter works better than MAF as the former imposes more weight on the most recent observation than the latter, thereby increasing the forecast capability of the model in which it is used. It has applications in finance and economics as well. It is remarkable to note that the main factor of the exponential smoothing method is the smoothing coefficient, which directly affects the accuracy of prediction.

We discuss one widely used exponential smoother, known as Holt exponential smoother, named after its discoverer.

This filter is recursive and simple, governed by the following iterative formula

$$y_t = \alpha x_t + (1 - \alpha)y_{t-1} \tag{5.1}$$

where α is an adjustable smoothing parameter. This filter is best suited for forecast studies, as it emphasizes the most recently occurring values and thus works equally well for all three components of a typical time series, that is, trend, periodic (seasonal) and random (irregular) patterns, for example, Holt-Winter forecasting.

- **Detrending high pass filter:**

This filter functions on the principle of bias removal, where we consider the difference between the original series and an appropriate low pass filtered series. The remainders obtained here are expected to retain the high pass element of the original series. For instance, the backward difference detrending filter

$$\Delta x = x_t - x_{t-1} \tag{5.2}$$

is twice the remainder left upon eliminating the MAF (2) low pass filtered trend from a time series.

Removing MAF (2) from a time series yields

$$\sum x_t - \sum \frac{1}{2} \sum_{k=-1}^{k=1} x_{t+k} = \frac{x_t - x_{t-1}}{2} = \frac{\Delta x}{2} \tag{5.3}$$

This filter removes stochastic trends and hence has applications in detrending time series. Commodity prices exhibit random behavior and therefore use this filter.

5.2.3 BAND PASS FILTER

This type of filter allows one band of frequency to pass through and rejects frequencies lower as well as higher than the chosen band. This band lies between low and high frequency ranges and rejects both low and high frequency components (Shumway et al., 2017). It comprises one pass band and two stop bands; it has two cut-off frequencies, one below and one above which the frequencies are attenuated. Mathematically, this may be thought of as a closed interval. Frequencies belonging to the interval are allowed; the rest are attenuated. Its graph can be seen in Figure 5.7.

Mankad et al. (2013) applied a 30- to 100-day band pass filter to thermocline depth (D20) to investigate intraseasonal oscillation in D20 and hence isolate its dominant peaks.

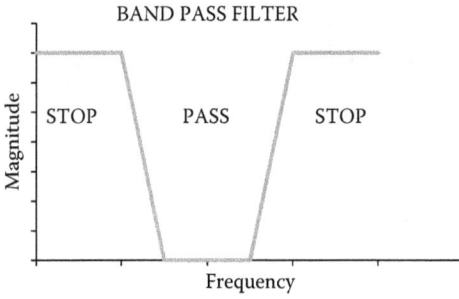

FIGURE 5.7 Band pass filter.

Further categories of high pass filters include a wide band pass and a narrow band pass filter, which are customized as per their requirement in designing electronic circuits.

To completely remove the biases of the parameter, Relative Humidity in Figure 5.4, a band pass filter of cut-off frequency 0.002–0.004 was applied to the entire span, and its result is shown in Figure 5.8. All the incorrect peaks (noise) randomly occurring over the temporal domain have been completely nullified, and a clear pattern is revealed.

Let us discuss the role of the band pass fast Fourier transform (FFT) filter in detail. In the band pass FFT filter, Fourier transformation of the input signal is first computed. Then, the transformed data is processed in the frequency domain. Finally, the altered frequencies are converted back to a signal in the time domain with inverse Fourier transformation. This filter removes the low-frequency and high-frequency tones. Specific pass band frequencies are given as input based on a requirement. The band pass FFT function can be defined as:

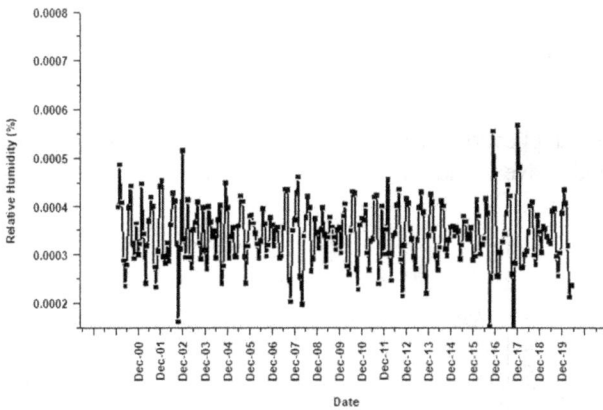

FIGURE 5.8 Relative Humidity (%) over the Himalayan region for the span 2000–2020 (ERA5-ECMWF).

$$w(f) = \begin{cases} 1, f \le f_{c1} \\ 1 - \dfrac{(f - f_{c1})^2}{(f_{c2} - f_{c1})^2}, \ f_{c1} < f < f_{c2} \\ 0, f \ge f_{c2} \end{cases} \tag{5.4}$$

where f_{c1} is the pass frequency and f_{c2} is the stop frequency.

High pass filters have extensive applications in wireless transmitters and receivers. They maintain the signal-to-noise ratio (S/N) for the receiver and limit the band width of the output signal in the transmitter. In optics, they are used in LIDARS, LASERS and contribute to audio signal processing by allowing a specific range of sound to pass through and attenuate the rest.

5.2.4 BAND STOP FILTER

This type of filter works complementary to the band pass filter, wherein it stops a particular range of frequencies and allows frequencies higher and lower than that range to pass through. It has one stop band and two pass bands. It is constructed using both low pass filter and high pass filter, both being connected in parallel. It possesses two cut-off frequencies; one occurs at low cut-off and another one at high cut-off. The mid frequencies are rejected, and the rest all are retained. It is also known as a band reject or band notch filter. Mathematically, we may look upon a band stop filter as the union of two semi-closed intervals, wherein the frequencies have a permissible range; everywhere else, or between these two intervals, the frequencies are prevented or stopped. Its graph is shown in Figure 5.9.

Band stop filters are widely used in audio signal processing, wherein a specific range of unwanted frequencies are removed and rest frequencies are retained. In communication systems, too, specific signals from a range of signals are refrained so as to have a better output. In public address systems and optical communication technologies, distortions or noise are eliminated using these filters.

5.3 EXTENDED APPLICATIONS OF VARIOUS CLASSES OF FILTERS IN REAL LIFE

Apart from the applications described in each of the sections above, filters and filtering methods have a wide range of applications in the field of medicine, too.

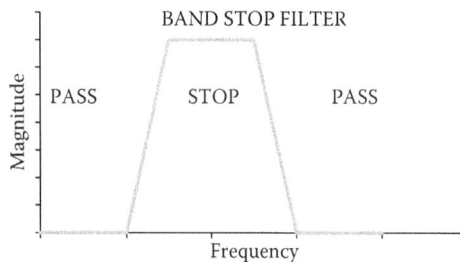

FIGURE 5.9   Band stop filter.

There is a branch called psychophysiology, wherein the data sets are in the form of time series. There are even periodic activities, which need to be monitored by a physician, in case any abnormal patterns are observed. The moving polynomial (obtained through best fit) filter is used to accurately describe rhythmic physiological processes (Porges & Bohrer, 1990). With the moving polynomial filter, it is possible to achieve a high level of smoothing in higher degree polynomials, without attenuating data features.

Exponential smoothers are used for forecast of weather data (predicting rainfall), which have both trend and seasonality (Marera, 2016). Further, various exponential smoothers have been compared to forecast the development trend of water environment security (Hong et al., 2010). This development trend would be used further to provide available scientific basis to guide water environment safety guarantee work. Band pass filters are applied in the fields of seismology, with sonars as well as in communication systems. Band stop filters are proven to be effective in musical instruments such as electric guitars. Other than this, they are useful in acoustic instruments like the mandolin. In biomedical instruments, the band stop filters play an active role by removing noise. In Digital Subscriber Line (DSL) Internet services, band stop filters remove interference on the line, thereby providing uninterrupted signals.

5.4 FUTURE SCOPE

Neither the types of filters, nor the applications are confined to what has been mentioned in this chapter. These filters have a wide spectrum of applications, covering almost every field of practical life. This chapter just drives the reader to explore more on time series analysis in general and filters in particular. Now, a lot of software is available for designing filters. For implementation and analyses purposes, one may use coding through MATLAB or Python, for the basic filters are inbuilt in the package.

REFERENCES

Hong, L., Yu, L., & Lin, L. (2010). Study on application of exponential smoothing method to water environment safety forecasting. *2010 International Conference on E-Product E-Service and E-Entertainment*, Henan (pp. 1–3).

Mankad, B. M., Sharma, R., Basu, S. K., & Pal, P. K. (2013). Intraseasonal thermocline variability in the equatorial Indian ocean. *Indian Journal of Geomarine Sciences*, *43*(1), 50–53.

Marera, D. H. (2016). An application of exponential smoothing methods to weather related data, A Research Report submitted to the Faculty of Science for the degree of Master of Science in the School of Statistics and Actuarial Science. University of Witwatersrand.

Montgomery, D. C., Jennings, C. L., & Kulahci, M. (2008). *Introduction to time series analysis and forecasting*. Wiley Interscience.

Peng, R. D. (2020). A very short course on time series analysis. https://bookdown.org/

Porges, S. W., & Bohrer, R. E. (1990). The analysis of periodic processes in psychophysiological research. In J. T. Cacioppo & L. G. Tassinary (Eds.), *Principles of psychophysiology: Physical, social, and inferential elements* (pp. 708–753). Cambridge University Press.

Shirahata, K., Yoshimoto, S., Tsuchihara, T., & Ishida S. (2016). Digital filters to eliminate or separate tidal components in groundwater observation time-series data. *Japan Agricultural Research Quarterly*, *50*(3), 241–252.

Shumway, R. H., & Stoffer, D. S. (2017). *Time series analysis and its applications: With R examples*. Springer Nature.

6 Deep Learning Approaches to Time Series Forecasting

Abhishek Sharma[1] and Sachin Kumar Jain[2]
[1]Power and Control, Dept. of ECE, PDPM-IIITDM Jabalpur
[2]Dept. of ECE, PDPM-IIITDM Jabalpur

6.1 INTRODUCTION

Time series forecasting is tremendously important to various fields like the stock market, environment, population growth, electric load, energy prices and economics. The key idea behind forecasting is extracting the useful features from historical data and predicting future timesteps. The features are used to build a mathematical model that can represent the actual process with reasonable accuracy. These models can broadly fall into three categories: first principle, statistical and neural network-based models.

The first principle or physics-based models are highly accurate models that are based on mathematical knowledge of the process. However, these models are not always feasible because we do not know the dynamics of the process or the process is too complex. The forecasting domain is generally very uncertain and involves nature or population behavior, so it is not a desirable category to choose.

Statistical methods involve the decomposition of the time series into three parts: trend, seasonality and irregular (Makridakis et al., 1997). The trend can be a deterministic or integrating type; it can be easily estimated using curve fitting for the deterministic type, or it can be differentiated for the integrating type (Tangirala, 2015). The seasonality can be determined by visualizing the autocorrelation plot for the series at various time lags. There can be multiple seasonalities present in a time series like daily, weekly, and yearly. The irregular or random part of the series is modeled using Box-Jenkins models (Box et al., 2016) like Autoregressive Moving Average (ARIMA) and Seasonal ARIMA (SARIMA). However, determining the model order is often subjective, and inaccurate estimation leads to large forecasting errors.

The neural network or learning-based methods do not require any assumption or knowledge about the process. The technique involves learning parameters known as weights, tuned by backpropagating the output error (Rumelhart et al., 1986). The Feed-Forward Neural (FNN) network can learn the representations in a given dataset but falls short in learning sequential representations. Recurrent neural network (RNN) architecture is adopted to overcome this limitation; it uses a weight-sharing

DOI: 10.1201/9781003102281-6

method and introduces a hidden state that distills the information in the series to a limited-sized vector. RNNs can learn time-stamped data, but they suffer from the problem of vanishing gradients, which prevents them from learning long sequences. Long short-term memory (LSTM) alleviates this problem by modifying the vanilla RNN discussed in the later section. Today, LSTM finds a vast number of applications in various time series forecasting and is the main highlight of this chapter.

This chapter discusses the recent advancements in time series forecasting using the context of electric load forecasting, so most of the literature cited is inspired by it. Electric load forecasting is one of the most widely acclaimed areas of time series prediction due to its importance in secure operation and control of power systems. The rest of the chapter is organized as follows: Section 6.2 presents state-of-the-art forecasting electric load techniques using deep learning (DL) methods. It provides a comprehensive literature review of the recent developments in this field. Section 6.3 discusses the DL architectures like RNNs and LSTMs and provides an insight into how LSTMs tackle the vanishing gradients problem. Section 6.4 provides a case study on France's electric load data; day-ahead forecast results are also presented. Section 6.5 concludes the chapter and gives the scope of further research.

6.2   STATE-OF-THE-ART FORECASTING

He (2017) proposed a DL approach for short-term load forecasting (STLF). The author used a convolutional neural network (CNN) to learn rich features and RNN to learn the historical load sequence dynamics. CNNs are well suited for image classification applications, but they are not useful for learning temporal behavior in a sequence. However, suppose CNN and RNN are used in conjunction. In that case, they can improve learning of representations, thus improving forecasting accuracy.

Jiao et al. (2018) proposed a method for STLF based on LSTM for non-residential loads using multiple correlated sequences. The multiple sequences are first classified using the k-means algorithm to determine the load consumption behavior. Spearman's correlation is used to find the time dependencies for different series. The method performed very well for regular patterned data, but LSTM's performance degraded drastically for irregular data. The addition of more features in this case also reduced the forecasting accuracy.

Residential load series possess a different challenge versus utility level or aggregated load. Each household can have uncertain load consumption behavior, making it highly volatile and very difficult to predict. Shi et al. (2018) proposed a two-staged method for residential load forecasting. The first stage pooled the individual household data, thus increasing the data volume and preventing overfitting for a small number of layers in the network. The second stage used the pooled data as input to a deep RNN for forecasting purposes. The model used was very naïve in form, and only a quantitative analysis was performed in this work, therefore leaving a margin for improvement in model architecture.

Kong et al. (2019) addressed the above problem and performed a density-based clustering of the household knowns with Density-Based Spatial Clustering of Application with Noise (DBSCAN) (Ester et al., 1996). DBSCAN does not require the number of clusters to initialize, and it also has the notion of outliers. Then, an

LSTM-based framework was adopted to forecast. The accuracy at the individual-level forecast was low; however, the forecast yield was much better after aggregation.

A detailed comparison of DL and traditional time series techniques for forecasting the building load is presented in Cai et al. (2019). Two methods are formulated based on gated CNN and LSTM for both recursive and direct multi-step forecasting. The forecasting accuracy improved for the DL approach as compared to the SARIMAX (traditional) method. The approach was computationally efficient as the DL architecture allows parallelization. The algorithm could be executed on a dedicated Graphical Processing Unit (GPU), thus reducing the main processor's burden.

Smyl (2020) presented a winning submission at the M4 competition; the paper proposed a hybrid algorithm of exponential smoothing and LSTMs. This hybrid method provides the advantages of both statistical and DL methods by complementing each other. The exponential methods can capture the seasonality and trend in the series. At the same time, LSTM can easily learn the non-linear patterns in the series. The ensembles of the models were built by considering top-performing models and building their pool. The final forecast would be the mean of these ensemble models.

Skomski et al. (2020) explored different questions that arise during the implementation of RNNs. The paper answered questions like dependence of forecasting accuracy on time resolution of data, the amount of training data required, forecast-horizon or lead time, the sensitivity of model w.r.t. hyper-parameters and generalization of the model. The author used a sequence-sequence model based on LSTM using an encoder-decoder mechanism to forecast. The drawback of the method is that it performs poorly beyond one-step prediction. A more extended input sequence is required to predict a longer output sequence.

The longer input-output sequence generally yields poor forecast accuracy, as mentioned above. To overcome this requires some modification to the existing mechanism. Du et al. (2020) used an attention mechanism (Sutskever et al., 2014), which focuses on the series' relevant portions and gives less priority to other parts. It does so by generating a temporal context vector. The method was based on bidirectional-LSTM with temporal attention, which facilitates the learning of long-term dependencies in multivariate time series.

6.3   DEEP LEARNING METHODS

Deep neural networks consist of a large number of hidden layers and neurons. This large number allows the network to learn the non-linearities present in the input, making it impossible for a shallow network to learn. This section presents DL techniques specially tailored for learning temporal data or sequences. The two standard architectures for this purpose are RNNs and LSTMs.

6.3.1 RECURRENT NEURAL NETWORKS (RNNs)

FNN networks can learn highly non-linear functions but are unsuitable for sequence learning. FNN networks do not consider temporal dependencies among the different timesteps of a time series. The expression for a simple RNN is given as:

$$\mathbf{h}_{(t)} = \sigma \left(\mathbf{W}_{xi}^T \mathbf{x}_{(t)} + \mathbf{W}_{hi}^T \mathbf{h}_{(t-1)} + \mathbf{b}_h \right)$$

$$\mathbf{y}_{(t)} = \sigma \left(\mathbf{W}_{hf}^T \mathbf{h}_{(t)} + \mathbf{b}_y \right)$$

$$(6.1)$$

where $\mathbf{h}_{(t)}$ is a hidden space vector that depends on the past state $\mathbf{h}_{(t-1)}$ and current input $\mathbf{x}_{(t)}$. The RNNs are trained using backpropagation through time (BPTT) in which the gradient flows through each timestep and updates the recurrent, input and output weight matrix, respectively. The weights are shared along every timestep. During BPTT, the gradient diminishes due to multiple differentiation and leads to negligible update in the weights at initial timesteps.

6.3.2 LONG SHORT-TERM MEMORY (LSTM)

LSTM mitigates the vanishing gradient issue inherent to vanilla RNNs and increases the span of the learning of long-term dependencies (Hochreiter & Schmidhuber, 1997). LSTM does so by introducing gates to conventional RNNs. The gates are namely input, forget and output, respectively (Gers & Cummins, 2000). They selectively read the information from past states and write it to the present state by processing it through the sigmoid function. The set of expressions for LSTM is given as:

$$\mathbf{i}_{(t)} = \sigma \left(\mathbf{W}_{xi}^T \mathbf{x}_{(t)} + \mathbf{W}_{hi}^T \mathbf{h}_{(t-1)} + \mathbf{b}_i \right)$$

$$\mathbf{f}_{(t)} = \sigma \left(\mathbf{W}_{xf}^T \mathbf{x}_{(t)} + \mathbf{W}_{hf}^T \mathbf{h}_{(t-1)} + \mathbf{b}_f \right)$$

$$\mathbf{o}_{(t)} = \sigma \left(\mathbf{W}_{xo}^T \mathbf{x}_{(t)} + \mathbf{W}_{ho}^T \mathbf{h}_{(t-1)} + \mathbf{b}_o \right)$$

$$(6.2)$$

$$\mathbf{g}_{(t)} = \tanh \left(\mathbf{W}_{xg}^T \mathbf{x}_{(t)} + \mathbf{W}_{hg}^T \mathbf{h}_{(t-1)} + \mathbf{b}_g \right)$$

$$\mathbf{c}_{(t)} = \mathbf{f}_{(t)} \otimes \mathbf{c}_{(t-1)} + \mathbf{i}_{(t)} \otimes \mathbf{g}_{(t)}$$

$$\mathbf{y}_{(t)} = \mathbf{h}_{(t)} = \mathbf{o}_{(t)} \otimes \tanh(\mathbf{c}_{(t)})$$

where $\mathbf{i}_{(t)}$, $\mathbf{f}_{(t)}$ $\mathbf{o}_{(t)}$ are the input, forget and output gates, respectively. The hidden state $\mathbf{h}_{(t)}$ is a vector which contains the distilled information from all the past inputs and states.

6.4   CASE STUDY

The data taken for the study is the electric load of France recorded by RTE from January 1, 2012 to December 31, 2017 with a 15-minute sampling rate. The data are

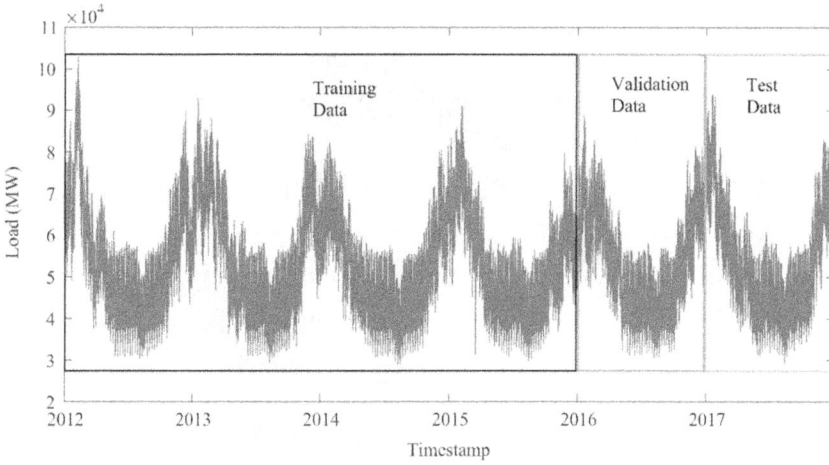

FIGURE 6.1   Training, validation, and test split of the electric load data of France (RTE).

split into three parts: training, validation and testing, as shown in Figure 6.1. Training data are used to train the model by giving it inputs and corresponding target value; error is calculated by the difference between the actual and predicted value. The error is then backpropagated to update the weights and repeated for a number of times, known as epochs.

The validation data is used to prevent the overfitting of the model. The training error is optimistic in nature, and due to a high number of epochs and a deep network configuration, the model tends to learn each point instead of generalizing the data. Validation error should be within given bounds and should not increase monotonically while the training error reduces.

Test data are used to determine the model accuracy. The metric used is the mean absolute percentage error (MAPE), given by:

$$MAPE = \frac{\sum_{i=1}^{N} \left| \frac{P_i - A_i}{A_i} \right|}{N} \times 100 \qquad (6.3)$$

The model was implemented on Keras (Chollet, 2015) using Tensorflow (Abadi et al., 2016) at the backend. The day ahead forecast results are illustrated in Figure 6.2. It shows the block-wise forecast average MAPE of 24 hours, that is, 96 blocks in the future.

The average MAPE for the whole duration compared with other methods like linear regression, non-linear regression and ARIMA models is shown in Table 6.1.

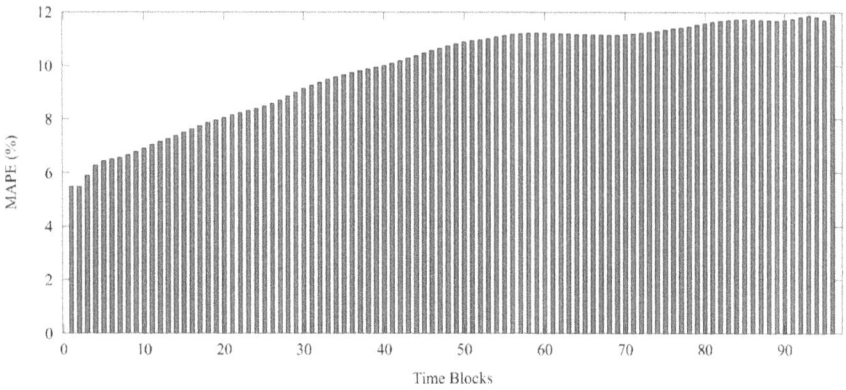

FIGURE 6.2 Average MAPE of 96 blocks for the duration between January 1, 2012 to December 31, 2017.

TABLE 6.1
Average MAPE for Test Data for the Year 2017

	Linear Reg.	Non-linear Reg.	ARIMA	LSTM
Avg. MAPE	19.92%	18.03%	14.20%	8.78%

6.5 CONCLUSION

Time series data contain non-linearities and uncertainties like consumer behavior, sensor noise, environmental effects and human body behavior like electroencephalograms and electrocardiograms. Thus, forecasting these series accurately becomes a big challenge and requires sophisticated algorithms. Various methods are available in the literature for time series forecasting. Statistical methods require few assumptions on the type of series and generally require more preprocessing like differencing, stationarity and model order identification. On the other hand, DL-based methods may require only data cleaning like outlier removal and filling-in of missing values.

DL-based methods are highly scalable and can be easily deployed on a local host machine. The inherent parallelization of the methods also allows training the model on multiple GPUs. The recent advancement in time series forecasting is shifting towards using machine learning, DL and hybrid (ML-statistical) methods. There are still several inadequacies left in these methods like the requirement of a high volume of data, large training time and less reproducibility of the results, which needs to be addressed for accurate and timely forecasts.

REFERENCES

Abadi, M., Barham, P., Chen, J., Chen, Z., Davis, A., Dean, J., Devin, M., Ghemawat, S., Irving, G., Isard, M., Kudlur, M., Levenberg, J., Monga, R., Moore, S., Murray, D. G.,

Steiner, B., Tucker, P., Vasudevan, V., Warden, P., … Zheng, X. (2016). TensorFlow: A system for large-scale machine learning. *Proceedings of the 12th USENIX Symposium on Operating Systems Design and Implementation, OSDI 2016.*

Box, G. E., Jenkins, G. M., Reinsel, G., & Ljung, G. M. (2016). *Time series analysis: Forecasting and control* (5th ed.). John Wiley & Sons.

Cai, M., Pipattanasomporn, M., & Rahman, S. (2019). Day-ahead building-level load forecasts using deep learning vs. traditional time-series techniques. *Applied Energy*, 236, 1078–1088. https://doi.org/10.1016/j.apenergy.2018.12.042

Chollet, F. (2015). *Keras documentation*. Keras.Io.

Du, S., Li, T., Yang, Y., & Horng, S. J. (2020). Multivariate time series forecasting via attention-based encoder–decoder framework. *Neurocomputing*, *388*, 269–279. https://doi.org/10.1016/j.neucom.2019.12.118

Ester, M., Kriegel, H.-P., Sander, J., & Xu, X. (1996). A density-based algorithm for discovering clusters in large spatial databases with noise. *Proceedings of the 2nd International Conference on Knowledge Discovery and Data Mining.*

Gers, F. A., & Cummins, F. (2000). Learning to forget: Continual prediction with LSTM. *Choice Reviews Online*, *12*(09), 2451–2471. https://doi.org/10.5860/choice.27-5238

He, W. (2017). Load forecasting via deep neural networks. *Procedia Computer Science*, *122*, 308–314. https://doi.org/10.1016/j.procs.2017.11.374

Hochreiter, S., & Schmidhuber, J. (1997). Long short-term memory. *Neural Computation*, *9*(8), 1735–1780. https://doi.org/10.1162/neco.1997.9.8.1735

Jiao, R., Zhang, T., Jiang, Y., & He, H. (2018). Short-term non-residential load forecasting based on multiple sequences LSTM recurrent neural network. *IEEE Access*, *6*, 59438–59448. https://doi.org/10.1109/ACCESS.2018.2873712

Kong, W., Dong, Z. Y., Jia, Y., Hill, D. J., Xu, Y., & Zhang, Y. (2019). Short-term residential load forecasting based on LSTM recurrent neural network. *IEEE Transactions on Smart Grid*, *10*(1), 841–851. https://doi.org/10.1109/TSG.2017.2753802

Makridakis, S., Wheelwright, S. C., & Hyndman, R. J. (1997). *Forecasting: Methods and principles*. Wiley.

Rumelhart, D. E., Hinton, G. E., & Williams, R. J. (1986). Learning representations by backpropagating errors. *Nature*, *323*, 533–536. https://doi.org/10.1016/j.measurement.2017.09.025

Shi, H., Xu, M., & Li, R. (2018). Deep learning for household load forecasting-A novel pooling deep RNN. *IEEE Transactions on Smart Grid*, *9*(5), 5271–5280. https://doi.org/10.1109/TSG.2017.2686012

Skomski, E., Lee, J. Y., Kim, W., Chandan, V., Katipamula, S., & Hutchinson, B. (2020). Sequence-to-sequence neural networks for short-term electrical load forecasting in commercial office buildings. *Energy and Buildings*, *226*, 110350. https://doi.org/10.1016/j.enbuild.2020.110350

Smyl, S. (2020). A hybrid method of exponential smoothing and recurrent neural networks for time series forecasting. *International Journal of Forecasting*, *36*(1), 75–85. https://doi.org/10.1016/j.ijforecast.2019.03.017

Sutskever, I., Vinyals, O., & Le, Q. V. (2014). Sequence to sequence learning with neural networks. *Advances in Neural Information Processing Systems*, *4*(January), 3104–3112.

Tangirala, A. K. (2015). *Principles of system identification*. CRC Press. https://doi.org/10.1201/9781315222509

7 ARFIMA and ARTFIMA Processes in Time Series with Applications

Kuldeep Kumar[1] and Priya Chaturvedi[2]
[1]Bond University
[2]Delhi School of Economics

7.1 INTRODUCTION

The Autoregressive Moving Average (ARMA) and Autoregressive Integrated Moving Average (ARIMA) models, introduced in the classic book by Box and Jenkins (1976), are able to capture short-range dependence. The dependence between time series observations of ARMA models decreases rapidly as the time lag increases. However, many economic time series depict long-range dependence, also called long-memory or long-range persistence. The autocorrelation function of such ARMA processes decays exponentially, and these processes fail to model the long-memory phenomenon present in a time series. The integrated processes have infinite lag memory, and the process reduces to a stationary process with short-range dependence after a finite number of differences. Obviously, the order of differencing for these processes is an integer.

The interesting feature of the Autoregressive Fractionally Integrated Moving Average (ARFIMA) process is that its autocorrelation functions decay much slower than exponential decay, which makes it a strong candidate for modeling a time series having a long memory. For time series possessing long memory, these ARFIMA processes provide an improved fit and better predictions in comparison to ARMA processes.

Let $\{Z_t, t = 1, ..., n\}$ be a realization of a time series with sample mean $\bar{Z}$ and variance S^2. The adjusted partial sums are defined by

$$S_t = \sum_{j=1}^{t} Z_t - t\bar{Z}, \quad t = 1, 2, ..., n$$

and the rescaled adjusted range by

$$R = S^{-1}\{\max(S_1, S_2, ..., S_n) - \min(S_1, S_2, ..., S_n)\}$$

DOI: 10.1201/9781003102281-7

If random variables $\{Z_j, j = 1, 2, \ldots\}$ are identically and independently distributed, one can easily verify that R is proportional to n^H with H = 0.5. However, while analyzing river flow time series, Hurst (1951) found the value of H to be around 0.73. The coefficient R is known as the "Hurst exponent" or "Hurst coefficient", and this phenomenon is known as the Hurst phenomenon, first observed by Hurst (1951, 1956) in geophysical time series. The observed value of H (0.73) led to the conclusion that it has been caused by long-range dependence or persistence in series.

The sharp contrast between the properties of persistent time series and those of the usual short-memory stationary time series has given rise to much controversy about the best way of modeling persistence. Several competing explanations of persistence have been proposed in the literature. Klemes (1974) and Potter (1976, 1979) have suggested that the Hurst phenomenon might be explained by non-stationarity. Boes and Salas (1978) have shown that sharp changes in the mean level can occur even in a process which is globally stationary. However, it is difficult to identify non-stationarity in a stochastic process through a single realization of the process.

In nutshell, long-range persistence in a time series is usually characterized by the following:

i. Instead of an exponential decaying autocorrelation function $\rho_j \sim \alpha^\beta$, $\alpha \in (0, 1)$, the autocorrelation function decays hyperbolically as the lag increases, i.e., $\rho_j \sim j^{-\beta}$, $\beta > 0$.

ii. The spectral density increases without bounds as the frequency tends to zero.

iii. The rescaled adjusted range (RAR) defined above.

iv. The Hurst coefficient behaves as a function $n^H (H > \frac{1}{2})$ of the sample size n.

7.2 MODELING PERSISTENCE: A PREVIEW

For a short-memory process, the RAR is considered to be a function of the sample size n. It increases eventually as $\sqrt{n} \to \infty$, though in finite samples this asymptotic relationship need not hold, and even for fairly large samples some processes can exhibit a Hurst exponent larger than $\frac{1}{2}$ (see Salas et al., 1979). Thus, some persistent series can be well fitted by short memory models. O'Connell (1971, 1974) has used ARMA(1,1) models for this purpose. Hipel and McLeod 1978 have fitted ARMA models to 23 geophysical time series. This means that a long-memory process can be modeled approximately by an ARMA model, though this would rarely yield the most parsimonious model.

Many researchers in hydrology have argued in favor of either fractional noise or an ARMA process for modeling persistence. According to Hosking (1984), however, these two approaches are not mutually exclusive but are the special cases of a general class of ARFIMA(p,d,q) models with $d \in R$, the real line. The fractionally integrated ARMA or ARFIMA process $\{Z_t\}$ is defined as

$$(1 - B)^d Z_t = X_t,$$
$$\Phi(B) X_t = \Theta(B) e_t$$

(7.1)

where $\Phi(B)$ is a polynomial of order p in backward shift operator "B", $\Theta(B)$ is polynomial of degree q in "B", and $\{e_t\}$ is a white noise process. For $d \in (0, \frac{1}{2})$, the resulting process is a stationary long-memory process with an autocorrelation structure similar to the fractional Gaussian noise process (Mandelbrot, 1971); with $d = 0$, one obtains the short-memory ARMA process, and when $d \in (0, \frac{1}{2})$ the process has a spectral density which vanishes at frequency zero and is called "anti-persistent". For a review of the earlier work on long-memory processes, fractional integration and their applications one may refer to Baillie (1996). Hurvich (2002) considered the fractional exponential (FEXP) model and developed forecasting procedures. He proposed algorithms for the computation of the coefficients in the infinite order ARMA representations of a FEXP process and approximating the autocovariances. Different chapters of the edited book by Robinson (2003) discuss various estimation methods and their theoretical properties along with empirical applications in several fields. Interesting discussion and comparison of ARFIMA programs concerning simulation, fractional order difference filter, estimation and forecast for various software platforms by Kai et al. (2017) may provide useful guidelines for selection of software.

7.3   ESTIMATION OF FRACTIONAL DIFFERENCING PARAMETER D

It has been observed that $H = d + \frac{1}{2}$. Hence, the estimate of d is obviously given by $\hat{d} = \hat{H} - \frac{1}{2}$. Several estimates of H using the principle have been suggested, but they are all highly biased, and their standard error is usually large. Keeping this in view, we generally do not recommend these methods based on the Hurst exponent.

Since spectral density provides another characterization of persistence, the estimator of d based on spectral density function can be constructed as discussed in detail in Singh and Kumar (1995). Given a realization $Z_1, Z_2, ..., Z_n$ of series $\{Z_t\}$, it is assumed that $\{Z_t\}$, when differenced d times, gives rise to a stationary series $\{X_t\}$ with a rational spectrum $f_X(\omega)$. Then, the spectrum of Z_t is given by

$$f_Z(\omega) = |1 - e^{-i\omega}|^{-2d} f_X(\omega)$$

(7.2)

Let

$$I_n(\omega, Z) = \frac{2}{n} \left| \sum Z_t e^{-i\omega t} \right|^2$$

be the periodogram at frequency ω. Janacek (1982) suggested an estimator of d, which fits a fractional exponential model to the log of periodogram ordinates at all available frequencies using a linear regression. The estimator of d is given by

$$\hat{d}_M = \left(S - \sum_{k=1}^{M} \frac{\hat{c}_k}{k} / \sum_{k=1}^{M} k^{-2} \right) \tag{7.3}$$

where

$S = \pi^{-1} \int_0^{\pi} W(\omega) ln \hat{f}_Z(\omega) d\omega,$

$W(\omega) = \sum_{k=1}^{\infty} \cos(k\omega)/k$

$\hat{c}_k = n^{-1} \sum_{p=1}^{n^*} ln I_n(b, Z) \cos(k\omega_p) + (2n^*)^{-1} ln I_n(0, Z) - \delta_n ln I_n(\pi, Z) \}$

with $n^* = (n - 1)/2$, $\delta_n = 1$ if n is even and 0 otherwise, $\omega_p = \frac{p}{n}$. The truncation point M is chosen large enough so that $c_k = 0$ for $k > M$. An alternative estimator of d at frequency $\omega_0 = m/n$, suggested by Parzen (1983), is

$$\hat{d}_k = \frac{1}{2} \left\{ k^{-1} \sum_{j=1}^{k} \ln \hat{f} \left\{ \frac{j+m}{n} \right\} - \ln \hat{f} \left\{ \frac{k+1+m}{n} \right\} \right\}$$

where $k = k_0$, $k_0 = 1, 2, \dots, n$; k_0 is a fairly large number and the spectrum $\hat{f}(\omega)$ can be estimated by a nonparametric kernel density estimator

$$\hat{f}(\omega) = \sum_{\nu=-\infty}^{\infty} k\left(\frac{\nu}{m} \right) \rho(\nu) \exp(-2\pi i \omega \nu) = \text{with } |\omega| < 0.5,$$

$\rho(\nu)$ is an estimator of the autocorrelation function(acf) at lag ν, and k(t) is the Parzen window defined by

$$k(t) = \begin{cases} 1 - 6t^2 + 6|t|^3, & |t| < 0.5 \\ 2(1 - |t|)^3, & 0.5 \le t \le 1 \\ 0, & otherwise \end{cases} \tag{7.4}$$

Geweke and Porter-Hudak (1983) considered the problem of estimating the parameter d in the general ARIMA(p,d,q) model. Let $Z_t \sim ARIMA(p, d, q)$ process and let $\nabla^d Z_t = X_t$, where $X_t \sim ARMA(p, d, q)$ process with spectrum

$$f_X(\omega) = \frac{\sigma^2 |\Theta(e^{-i\omega})|^2}{|\Phi(e^{-i\omega})|^2}$$

which is bounded and bounded away from zero and continuous on $[-\pi, \pi]$. Then,

$$f_Z(\omega) = |1 - e^{-i\omega}|^{-2d} f_X(\omega) \tag{7.5}$$

Taking log and adding $lnf_X(\omega) = lnf_X(0) - 2d \ln|1 - e^{-i\omega}| + \ln\left[\dfrac{f_X(\omega)}{f_X(0)}\right]$

$$(7.6)$$

Furthermore, replacing ω in equation (7.6) by the Fourier frequency $\omega_j = 2\pi j/n$ and adding $lnI_n(\omega_j)$ to both sides of equation (7.6) one obtains

$$lnI_n(\omega_j) = lnf_X(0) - 2dln\left|1 - \exp\left(-i\omega_j\right)\right| + \ln\left(\dfrac{I_n(\omega_j)}{f_Z(\omega_j)}\right) + \ln\left(\dfrac{f_X(\omega_j)}{f_X(0)}\right) \quad (7.7)$$

Omitting the last term on the RHS of equation (7.7), which is negligible and re-writing equation (7.7) in the regression form, we have

$$y_j = \alpha + \beta v_j + \varepsilon_j \,, \quad j = 1,\, 2,\, \ldots, m \qquad (7.8)$$

where $y_j = lnI_n(\omega_j)$, $v_j = ln|1 - \exp(-i\omega_j)|^2$, $\varepsilon_j = lnI_n(\omega_j)/f_Z(\omega_j)$, $\alpha = lnf_X(0)$ and $\beta = -d$, m is a function of such that $\frac{m}{n} \to 0$ as $n \to \infty$. The least square estimator of d is then given by

$$\hat{d} = -\sum_{i=1}^{m}(y_j - \bar{y})/\sum_{i=1}^{m}(v_i - \bar{v})^2 \qquad (7.9)$$

Hosking (1984) discussed the maximum likelihood method, providing efficient estimators of parameters of fractionally differenced models. However, the details of Hosking's procedure are not discussed here.

7.4   APPLICATION OF ARIMA TO CRUDE OIL DATA

This section provides data application of the ARFIMA (p, d, q) time series model, using the "arfima" package in R. We collected the adjusted crude oil price (dollars per barrel) data for the time period January 1, 2011 to January 1, 2019 (weekly series) from Yahoo! Finance. The ACF plot in Figure 7.1 shows that ACF decays slowly and not exponentially.

To check stationarity of the series, we used the Augmented Dickey Fuller (ADF) test. The ADF test examines the null hypothesis that a time series Y_t is stationary against the alternative that it is non-stationary. The p-value turned out to be 0.066, which is greater than 5%. Thus, we can conclude that the series is not stationary.

We use the Hurst exponent (H) (using function available in "pracma" package in R) to test the presence of long memory in the data. The value of H turned out to be 0.8652, indicating that the data has a long-memory structure since $0.5 < H < 1$.

The long memory parameter d is estimated using the package "fracdiff" in R. Estimated value of the parameter, its asymptotic deviation value and regression standard deviation values are shown in Table 7.1.

ACF of Adjusted Crude Oil Price

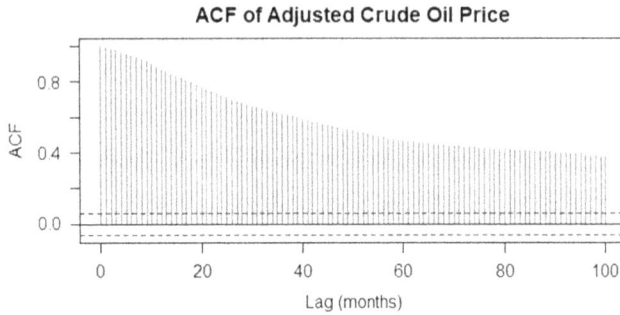

FIGURE 7.1   ACF plot of adjusted crude oil price.

TABLE 7.1

Estimated Value of the Parameter, Its Asymptotic Deviation Value and Regression Standard Deviation Value

Coefficient	Estimate	Asymptotic Std. Deviation	Std. Error Deviation
d	0.9340	0.1372	0.1536

TABLE 7.2

Best Models According to AIC and BIC Criterion

	Best	Second Best	Third Best	Fourth Best
AIC Models	ARFIMA(0,0,0)	ARFIMA(0,0,0)	ARIMA(0,0,1)	ARIMA(0,0,0)
AIC	5103.51	5114.64	5115.27	5121.39
p(AIC)	1.000	0.004	0.003	0.000
BIC Models	ARFIMA(0,0,0)	ARIMA(1,0,0)	ARIMA(0,0,1)	ARIMA(0,0,0)
BIC	5118.20	5129.33	5129.96	5131.18
p(BIC)	1.000	0.004	0.003	0.002

After conducting an ADF test for the fractionally differenced series, we inferred that the series is stationary and can be used for further analysis.

We used the best models function in the ARTFIMA package in R to identify the best ARFIMA model (Table 7.2).

From the table, we saw that the best model according to AIC is ARFIMA(0,0,0), and according to BIC, it is also ARFIMA(0,0,0). We fitted ARFIMA(0,0.934,0) using the ARFIMA package using the Whittle estimator (Table 7.3).

The AIC and BIC of the fitted model were 5103.61and 5118.3, respectively.

TABLE 7.3

Estimates of the Fitted Model

	Estimate	S.E. (Estimate)
Mean	−0.0561	0.1446
d	0.0990	0.0171

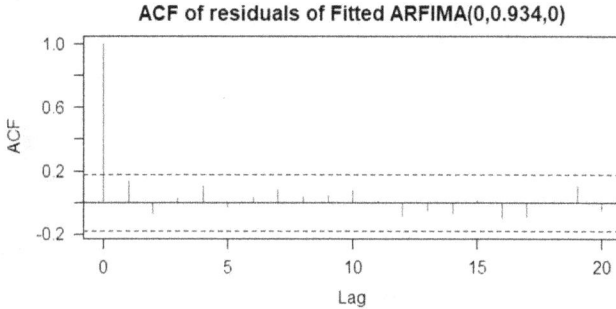

FIGURE 7.2 ACF plot of residuals of fitted ARFIMA(0,0.934,0).

The residual ACF of the fitted ARFIMA(0,0.934,0) model of the fractionally differenced series is shown in Figure 7.2. The plot shows that there is no serial correlation observed in the residuals of the series.

7.5 ARTFIMA PROCESSES

For modeling semi-long-range dependence in a time series, Giraitis et al. (2000) introduced the tempered fractionally integrated (TFI) time series model. The interesting characteristic of ACF of the TFI model is that its ACF echoes long-term dependence for a number of lags, but ultimately decays exponentially due to the tempering parameter. Meerschaert et al. (2014) introduced the Autoregressive Tempered Fractionally Integrated Moving Average (ARTFIMA) time series model as a generalization of the TFI model. Sabzikar et al. (2015) demonstrated that tempered fractional Gaussian noise provides a useful model for wind speed data.

We define the tempered fractional difference (TFD) operator as

$$
\begin{aligned}
\nabla^{d,\lambda} Z_t &= (I - e^{-\lambda}B)^d Z_t \\
&= \sum_{j=0}^{\infty} \omega_j^{d,\lambda} Z_{t-j}, \quad d > 0, \ \lambda > 0
\end{aligned}
\tag{7.10}
$$

with

$$\omega_j^{d,\lambda} \equiv (-1)^j \binom{d}{j} e^{-\lambda j}$$

and

$$\binom{d}{j} = \frac{\Gamma(1+d)}{j!\Gamma(1+d-j)}$$

Here $\Gamma(d) = \int_0^\infty e^{-x} x^{d-1} dx$ is the usual gamma function. It is interesting to note that for $\lambda = 0$ the process (5.1) reduces to the ARFIMA operator

$$\nabla^{d,0} Z_t = (I - B)^d Z_t$$
$$= \Sigma_{j=0}^\infty (-1)^j \binom{d}{j} Z_{t-j}. \tag{7.11}$$

A time series $\{X_t\}$ is said to follow a *ARTFIMA* $(p; d; \lambda; q)$ process if it can be represented as

$$\Phi(B) \nabla^{d,\lambda} Z_t = \Theta(B) e_t \tag{7.12}$$

Thus, $X_t = \nabla^{d,\lambda} Z_t$ follows a *ARMA* (p, q) process.

Since the ACF of ARFIMA processes is not summable, it often becomes difficult to analyze statistical properties of these processes. The advantage with ARTFIMA processes is that these processes have summable ACF, making it convenient to investigate statistical properties. Still, one can choose the tempering parameter arbitrarily small so that the process resembles the ARFIMA process. In this way, one can fit it to data which are usually fitted using ARFIMA processes and also keep the process mathematically more tractable than the ARFIMA process. The ARTFIMA model provides a better fit than the ARFIMA processes when the power spectrum is bounded, even when the frequency approaches zero, a case when the ARFIMA process fails to provide an appropriate fit as its power spectrum tends to infinity as frequency approaches zero. One such example is turbulence data; see Meerschaert et al. (2014) for detailed discussion. Another advantage of ARTFIMA processes is that these models are stationary for any order of tempered fractional differencing. It is unlike the other models used for modeling long-range dependence such as the ARFIMA process or fractional Brownian motion, which have stationary increments but don't have stationary series for all values of the Hurst index. Thus, ARTFIMA processes should be preferred for modeling stationary time series having long memory.

Sabzikar et al. (2019) provided parameter estimation methods for the ARTFIMA model and implemented it using a new R package called "artfima". They proposed the Whittle estimators for estimating the parameters of the ARTFIMA(p,d,λ,q) model and established the consistency and asymptotic normality of the maximum likelihood and for the parameters. For deriving the Whittle estimators, we can write the spectral density of ARTFIMA(p,d,λ,q) as

$$f_X(\nu; \theta) = \frac{\sigma^2}{2\pi} \frac{|\Theta(e^{-i\nu})|^2}{|\Phi(e^{-i\nu})|^2} |1 - e^{-(\lambda+i\nu)}|^{-2d} : = \frac{\sigma^2}{2\pi} K(\nu, \theta); \ \nu \in (-\pi, \pi) \quad (7.13)$$

with $\theta = \left(\phi_1, ..., \phi_p, \theta_1, ..., \theta_q, d, \lambda\right)$. For N observations $X = (X_1, ..., X_N)$ from ARTFIMA(p,d,λ,q), the estimated spectral density (periodogram) is

$$I_X(\nu): = \frac{1}{2\pi N} \left| \sum_{t=1}^{N} X_t e^{-it\nu} \right|^2 \quad (7.14)$$

Let

$$Q(X, \theta): = \int_{-\pi}^{\pi} \frac{I_X(\nu)}{K(\nu, \theta)} d\nu$$

$$D_N(X, \sigma, \theta): = \frac{1}{2\sigma^2} Q(X, \theta) + \ln\sigma$$

Then, the Whittle estimators of θ and σ^2 are defined as

$$(\hat{\theta}_N, \hat{\sigma}_N): = \text{arg min} \{D_N(X, \sigma, \theta)\}$$

We can easily verify that $\hat{\theta}_N = \text{arg min} \{Q_X(\theta)\}$, $\hat{\sigma}_N^2 = Q_X(\hat{\theta}_N)$.

They have shown that, for large N, the Whittle estimator is approximately equal to the maximum likelihood estimator and much easier to compute. Hence, in practice, it provides a useful approximation. They also derived the asymptotic covariance matrix of the estimator.

7.6 APPLICATION OF ARTFIMA TO CRUDE OIL PRICES

We used the best_glp_models function available in the ARTFIMA package in R to find the best ARTFIMA model that fits the fractionally differenced crude oil price. The best ARTFIMA model is ARTFIMA(0,0,2) with AIC 5098.069.

We fitted an ARTFIMA model with p = 0, q = 2 and obtained Whittle estimators for parameters using the artfima package. The fitted parameters are d = 0.6165 (0.1621), λ = 0.095(0.02746971), θ_1 = 0.49308137 (0.14995778), θ_2 = 0.108 (0.0293). The term in the parentheses is the standard error.

Figure 7.3 shows that the resulting model spectral density provides a reasonable fit to the periodogram.

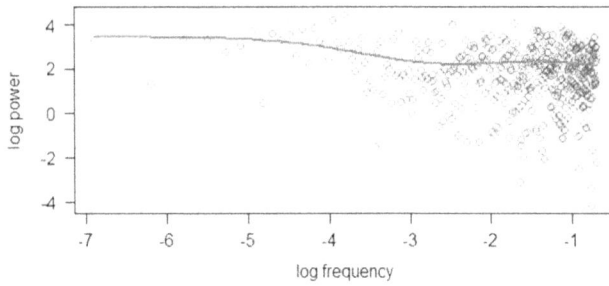

FIGURE 7.3   Spectral density of fractionally differenced crude oil price (circles in blue), fitted ARTFIMA spectrum (red line).

7.7   CONCLUDING REMARKS

The purpose of this study was to fit ARFIMA and ARTFIMA models to the fractionally differenced adjusted crude oil price. In Section 7.1, we defined briefly the long-memory (or persistent) series, and in Section 7.2, we gave a brief account of relevant references. Section 7.3 gave a summary of the estimation procedures, mainly for the differencing parameter d. In view of the poor performance of some of the existing methods and the algebraic complexities of almost all available methods, we suggested a heuristic approach to the estimation problem in Section 7.4. The method was conjectured on geometrical grounds.

The function best_models in the package allows one to pick a best model for the data. The function best_glp_models helps to find the best ARMA/ARFIMA/ARTFIMA model with appropriate parameters.

The fitted ARTFIMA model has a lower AIC compared to the fitted ARFIMA model and hence provides a better fit to the differenced series.

REFERENCES

Baillie, R. T. (1996). Long memory processes and fractional integration in econometrics. *Journal on Econometrics*, *73*, 5–59.

Boes, D. C., & Salas, J. D. (1978). Nonstationarity of the mean and the Hurst phenomenon. *Water Resources Research*, *14*, 135–143.

Box, G. E. P., & Jenkins, G. M. (1976). *Time Series Analysis forecasting and control.* Holden-Day.

Geweke, J. F., & Porter-Hudak, S. (1983). The estimation and application of long-memory time series models. *Journal of Time Series Analysis*, *4*, 221–238.

Giraitis, L., Kokoszka P., & Leipus R. (2000). Stationary ARCH models: Dependence structure and central limit theorem. *Econometric Theory*, *16*, 3–22.

Hipel, K. W., & McLeod, A. E. (1978). Preservation of the rescaled adjusted range, 2, simulation studies using Box-Jenkins models. *Water Resources Research*, *14*, 509–516.

Hosking, J. R. M. (1984). Modelling persistence in hydrological time series using fractional differencing. *Water Resources Research*, *20*, 1898–1908.

Hurst, H. E. (1951). Long-term storage capacity of reservoirs. *Transactions of the American Society of Civil Engineers*, *116*, 770–799.

Hurst, H. E. (1956). Methods of using long-term storage in reservoirs. *Proceedings of the Institution of Civil Engineers*, *1*, 519–543.

Hurvich, C. M. (2002). Multi-step forecasting of long memory series using fractional exponential models. *International Journal of Forecasting*, *18*, 167–179.

Janacek, G. J. (1982). Determining the degree of differencing for time series via the log spectrum. *Journal of Time Series Analysis*, *3*, 179–183.

Kai, L., Chen, Y. Q., & Zhang, X. (2017). *Axioms*, *6*, 16. https://doi.org/10.3390/axioms6020016.

Klemes, V. (1974). The Hurst phenomenon- a puzzle? *Water Resources Research*, *10*, 675–688.

Mandelbrot, B. B. (1971). A fast fractional Gaussian noise generator. *Water Resources Research*, *7*(3), 543–553.

Meerschaert, M. M., Sabzikar, F., Phanikumar, M. S., & Zeleke, A. (2014). Tempered fractional time series model for turbulence in geophysical flows. *Journal of Statistical Mechanics* , 2014, P09023.

O'Connell, P. E. (1971). A simple stochastic modelling of Hurst law. In *Mathematical models in hydrology* [Symposium]. Warsaw (Vol. 1, pp. 169–187).

Parzen, E. (1983). Time series model identification by estimating information, memory and quartiles [Technical Report, Texas A & M University].

Potter, K. W. (1976). Evidence for nonstationarity as a physical explanation of the Hurst phenomenon. *Water Resources Research*, *12*, 1047–1052.

Potter, K. W. (1979). Annual precipitation in the northeast United States: Long memory, short memory or no memory. *Water Resources Research*, *15*, 340–346.

Robinson, P. M. (2003). *Time Series with Long Memory*. Oxford University Press.

Sabzikar, F., Meerschaert, M. M., & Chen, J. (2015). Tempered fractional calculus. *Journal of Computational Physics*, *293*(2015), 14–28.

Sabzikar, F., Meerschaert, M. M., & Mcleod, I. (2019). Parameter estimation for ARTFIMA time series. *Journal of Statistical Planning and Inference*, *200*, 129–145.

Salas, J. D. D., Boes, C., Yevjevich, V., & Pegram, G. G. S. (1979). Hurst phenomenon as pre-asymptotic behaviour. *Journal of Hydrology*, *44*, 1–15.

Singh, N., & Kumar, K. (1995). *Fractional modelling of sales of an item* [Working paper], Australia Department of Statistics, Monash University.

APPENDIX

R code for fitting ARFIMA Process

```
fitarfima<- artfima(y, glp = c("ARFIMA"), arimaOrder = c(0, 0, 0), likAlg = c
   ("Whittle"))
```

R code for fitting ARTFIMA Process

```
fitartfima<-artfima(y, glp = c("ARTFIMA"), arimaOrder = c(0, 0, 2), likAlg = c
   ("Whittle"))
plot(fitartfima, which = "all", subQ = FALSE, mainQ = FALSE)
```

8 Comparative Study of Time Series Forecasting Models for COVID-19 Cases in India

Vasundhara Mahajan, Rishil Shah, Dharmik Bhatt, Darsh Patel, Prachi Agrawal, and Lalit Tak
Department of Electrical Engineering, Sardar Vallabhbhai National Institute of Technology, Surat, Gujarat, India – 395007,

8.1 INTRODUCTION

We review the spread of COVID-19, its impact on India and clinical developments in this section.

8.1.1 General Information about COVID-19

Coronavirus, or COVID-19, is a contagious respiratory disease caused by SARS-CoV-2, a specific type of virus responsible for the current pandemic. This SARS-CoV-2 virus belongs to the same family as the MERS-CoV virus, responsible for the MERS outbreak in the past. Taxonomically speaking, this virus is a strain of SARSr-CoV. Its origins have a close genetic resemblance to bat coronaviruses, implying it may trace its roots to a bat-borne virus (Wikipedia contributors, 2020). The first case of this virus was reported in the Wuhan district of China in December 2019, which is also responsible for the name – Novel CoronaVirus – 19 (Geneva: WHO, 2020). The World Health Organization (WHO) declared the COVID-19 outbreak a pandemic on March 11, 2020 (Wikipedia contributors, 2020). As of November 4, 2020, the total number of cases worldwide is 47.5 million, with 31.6 million recoveries and 1.21 million deaths (Geneva: WHO, 2020).

Person to person is the primary transmission mode of the COVID-19 virus. Whenever a person comes in contact with the virus, he/she becomes a potential carrier of that virus irrespective of whether the person is symptomatic or asymptomatic. As this is a respiratory disease, transmission primarily occurs when an infected person coughs or sneezes and the contaminated air droplets are released into the environment. When these droplets reach the mouth, ears or eyes of a healthy person, he/she gets infected, and the chain continues. Studies report that the virus can survive on surfaces like doorknobs, banknotes, plastics, copper, stainless steel, etc., for as

DOI: 10.1201/9781003102281-8

long as 72 hours. If a person touches any of these surfaces and subsequently touches his/her mouth, eyes or ears, then the person is likely to get infected.

To curb the increasing number of cases and reduce the transmission rate, the WHO issued strict guidelines and advisories to prevent over-burdening the health care infrastructure in highly populated countries. Countries like India, England, and New Zealand went under complete lockdown during the initial period to control the rate of transmission. With its strict lockdown in the early days of the virus spread, New Zealand was the first country to successfully control the transmission of COVID-19, with the total infection cases as low as 1,800–1,900, while other countries still struggle to control the spread. The United States, which is the worst affected country to date, with approximately 90,000+ reported cases and deaths in the range of 500–1,000 per day, did not initially follow the guidelines laid down by the WHO, nor did the government impose any kind of lockdown—partial or complete whatsoever—which led to a catastrophic outcome. The United States currently accounts for over 10 million reported cases.

8.1.2 Impact on India

The first COVID-19 case in India was reported on January 30, 2020. Currently, India is the worst-hit country in Asia and the second worst-hit country worldwide. The impact of COVID-19 on India has been largely disruptive. In a country like India, which was already experiencing an economic slowdown, the pandemic has made things far worse (Wikipedia contributors, 2020). The primary reason for the blow to the economy is the nation-wide lockdown imposed in March by the government of India. However, the steps were appreciated by the WHO, as it was a necessary move to curb the spread of the virus. The Oxford COVID-19 government response tracker assigned a 100 score on its stringency index for effective emergency response. As of November 4, 2020, the total number of confirmed cases in India is 8.31 million with 124,000 deaths (COVID-19 India Org Data Operations Group, 2020). The recovery rate saw an increase from 11.4% in April 2020 (Economic Times, 2020) to 92.09% in November 2020 (Times of India, 2020).

Maharashtra is the worst-hit state in India, with 1.6 million cases reported, followed by Karnataka, Andhra Pradesh and Tamil Nadu (COVIDIndia.org, 2020). Dharavi, which is Asia's largest slum, with a population density of 277,136 per km^2, is one of the most densely populated places in the entire world. Despite being so densely populated and consequently having the potential of being the largest COVID-19 cluster in the country or even in Asia for that matter, the Maharashtra government has successfully contained the transmission of COVID-19 with its new and innovative response policy. Dharavi successfully flattened the COVID-19 transmission curve in just two months with a policy called "chasing the virus", rather than waiting for people to report it. The commendable feat was achieved by actively tracing, tracking, testing and treating potential carriers of the virus (Golechha, 2020).

On the other hand, Kerala was able to control the situation by showing early and prompt preparedness (WHO India, 2020). The state declared a health emergency as soon as the first case was reported, and screening and surveillance of all the passengers coming from China was the initial response. Further steps, like systematic investments in health care in the form of setting up dedicated COVID-19 hospitals and trained teams to treat the increasing positive cases, were put into immediate action (WHO India, 2020). Intensive testing is being conducted by the government of India with the total number of tests done by November 4, 2020 at 1.12 billion (COVIDIndia.org, 2020).

8.1.3 IMPACT ON DAILY ACTIVITIES

The major impact of the lockdown involved the complete closure of all the commercial establishments like restaurants, theatres and malls. Schools and colleges had to move their functioning online, and a majority of examinations were postponed. The effect on the stock market was huge. The market suffered a huge crash on the day the WHO publicly acknowledged COVID-19 as a pandemic, with BSE Sensex dropping by 8.18% (2,919 pts) and NIFTY by 9% (950 pts), respectively. All religious buildings, including places of worship, were closed, and travel was virtually stopped, with significant restrictions on inter- and intra-state travel via any mode—road, rail or air (COVID W., 2019).

The MHA issued guidelines for unlocking India in a phased manner with five unlock phases in mind. Each phase of the unlock witnessed the removal of certain restrictions, ranging from UNLOCK 1.0—reopening religious places, malls and hotels, allowing inter-state travel, etc.—to UNLOCK 5.0—reopening schools (final decision to be taken by the state/UT government), cinema halls with 50% seating capacity and swimming pools for the training of sportspersons. Restrictions still remain in place in containment zones without any relaxation (Wikipedia contributors, 2020).

8.1.4 VACCINE DEVELOPMENTS

Researchers and scientists around the world are working tirelessly to find a vaccine for the novel coronavirus. A vaccine by the Gamaleya Research Institute in Moscow (Ray, 2020), which goes by the name Sputnik V, was approved by the Ministry of Health of the Russian Federation. This vaccine is yet to enter phase 3 clinical trials, which is why considerable concerns were raised about its safety.

Currently, there are around 51 vaccine candidates in development, out of which eight are in phase 3 of clinical trials as of November 3, 2020, while two are in phase 2/3. All the other candidates are either in phase 1, phase 2 or the pre-clinical and early research phase (Craven, 2020).

8.1.5 RESEARCH EFFORT FOR COVID-19

Currently, the COVID-19 pandemic is one of the most severe issues affecting the contemporary world because of its extremely detrimental effects on public health (Toğaçar et al., 2020). It has taken a catastrophic toll on disadvantaged groups, particularly the elderly and individuals with chronic illnesses, like those with asthma. As the world continually fights against COVID-19 by taking various precautionary measures, the scientific community is making significant efforts to expedite vaccination development and model the growth of the virus. The ability to identify the pace of disease spreads is critical for the battle against the pandemic. Knowledge of the extent of transmission at any particular moment has the ability to assist policymakers in public health preparation and policy-making to address the effects of the pandemic (Klompas, 2020; Preiser et al., 2020).

In recent months, numerous studies have been conducted by researchers across the globe to predict the possible impact of COVID-19 in different regions, to support timely decisions for arresting the spread of the virus. Various modeling, forecasting and statistical methods have been applied to discern and control the outbreak in different countries. For predicting the mortality rate due to the pandemic in China (Wang et al., 2020), researchers employed the person information based algorithm. The study estimated the total death rate to be around 13% in Hubei and Wuhan and between 0.75% and 3% in the remaining provinces of China. The combined analysis and forecasting using a susceptible infected recovered deaths (SIRD) model for China, Italy and France allowed researchers to roughly estimate the number of ventilation units required to sustain the increasing number of patients in Italy (Fanelli & Piazza, 2020). In (Jia et al., 2020), the Logistic, Bertalanffy and Gompertz models have been employed for the analysis and prediction of cases. Compared with the other two methods, the Logistic model demonstrated accurate predictive performance. However, these three models need adequate data and have limited applicability to only some outbreak phases. Similarly, various other conventional time series forecasting models have been used to predict possible COVID-19 cases in China and a few other nations (Kucharski et al., 2020, Wu et al., 2020).

Focusing on deep learning methods, Zeroual et al. (2020) presented a comparative study of recurrent neural network (RNN) and its variants like long short-term memory (LSTM) for estimating the confirmed and recovered cases in six different countries. The experimental results established the extended capability of deep learning models to capture non-linearity in the process and their versatility in time-dependent data modeling. Particularly, LSTM models have been shown to produce more accurate results in a number of studies. Chimmula and Zhang (2020) used only the LSTM model for predicting an end date for the epidemic in Canada. Their model obtained an accuracy of 93.4% and 92.67% in the short-term and long-term scenarios, respectively. In Arora et al. (2020), researchers employed three variants of the LSTM model for forecasting the daily and weekly trend of COVD-19 cases in different states in India. Using the bidirectional LSTM (Bi-LSTM) model, they were able to obtain short-term predictions with an error margin of only 3%.

The chapter describes and compares deep learning and machine learning models for predicting the COVID-19 case trends in ten of the worst-affected states in India for a time horizon of 15 days. The chapter explores six varieties of prediction models given as:

    A. Three predictive machine learning models
        i. Support vector regression (SVR)

    ii. Vector autoregression (VAR)

    iii. Polynomial regression (PR)

B. Three deep learning models

    i. Recurrent neural network (RNN),

    ii. Long short-term memory (LSTM),

    iii. Gated recurrent units (GRU)

The organization of the work is as follows: Section 8.2 discusses traditional and recent time series forecasting models and their applications in the medical domain. In Section 8.3, we give the mathematical background of all the methods to explain RNN variants. In Section 8.4, we present a descriptive analysis of the existing Indian data in a state-wise manner, and we discuss the spread of the virus in different regions. In Section 8.5, we explain the model selection and performance evaluation using the proposed models and discuss the trend of COVID-19 cases in the upcoming days. Finally, in Section 8.6 we present the conclusions derived from the study.

8.2 RELATED WORK

This section is dedicated to literature review and includes:

1. The evolution of time series forecasting models from traditional mathematical approaches to newer deep learning models.
2. Older research studies with machine learning models and statistical methods for forecasting the trends of infectious diseases.
3. Recent studies that have employed various forecasting models to predict the COVID-19 outbreak-related statistics with deaths and recoveries.

Time series forecasting is a domain in which prior measurements of a random variable are evaluated to create a model for capturing the basic trends and their connections. This model estimates the future values of the variable. This approach is particularly effective in cases: (1) when there is no explanatory model that can establish an accurate link between the prediction variable and other explanatory variables; and (2) when little to no information about the distribution or process is available. In the past few decades, a lot of research has been directed towards developing new time series forecasting models and improving existing ones (Papastefanopoulos et al., 2020).

In earlier years, traditional forecasting techniques used mathematical formulas. These simple approaches have been further enhanced through numerous developments in various tools and automation systems. Past applications include modeling of seasonal leptospirosis and its relationship with temperature and rainfall (Chadsuthi et al., 2012) and analysis of correlations between monthly count of *Plasmodium falciparum* cases and the El Niño Southern Oscillation (ENSO) (Hanf et al., 2011). The modeling of viruses in cyclic patterns or recurring cycles such as seasonal influenza has often been accomplished by similar methods and has been used to predict outbreaks by time series forecasting. In Song et al. (2016), an Autoregressive Integrated Moving Average (ARIMA) model is used to predict the monthly cases of influenza in China for 2012, while in Yin et al. (2020), a time series forecasting model (Tempel) is presented for "influenza A" virus mutation expectations.

In recent years, the medical community is generating enormous amounts of data on a daily basis. To leverage the data and gain insights through it, all kinds of machine and deep learning models are being used. Tasks such as segmentation, localization, classification and prediction are being achieved with state-of-the-art deep learning models with almost human-level accuracy. Few deep learning techniques have earned their place in the medical domain for analyzing certain types of data. Convolution networks are mainly responsible for the classification and segmentation tasks, whereas the RNN and its variants (LSTM, Bi-LSTM, stacked LSTM, GRU, etc.) are responsible for the processing and analysis of sequential data, time series forecasting and predictive analysis.

Accurate modeling and forecasting of COVID-19 cases has proved to be invaluable for optimizing the use of scarce health care resources and formulating effective management strategies for infected patients in highly populated countries like India. In this context, machine learning and deep learning forecasting models have recently appeared in a number of studies and applications, producing accurate predictions in a number of scenarios. Rustam et al. (2020) employed four commonly used machine learning models for forecasting the number of new cases, recoveries and deaths for a period of ten days. Evaluation of all four models using different performance metrics established exponential smoothing (ES) as the best model considering the nature and size of the data set. Along similar lines, Tuli et al. (2020) provided a comprehensive framework for modeling and estimating the future dispersion of the epidemic in major countries using cloud computing and machine learning. Saba and Elsheikh (2020) used the Autoregressive Integrated Moving Average (ARIMA) and the Non-Linear Autoregressive Artificial Neural Networks (NARANN) to predict the increasing COVID-19 cases in Egypt. With the absolute percentage error of both the proposed models within 5%, the resulting predictions were useful in devising short-term strategies to arrest the spread in the region.

8.3 METHODS

This section describes the predictive machine and deep learning models.

The predictive machine models are: (1) support vector regression (SVR), (2) polynomial regression (PR) and (3) vector autoregression (VAR).

The deep learning models are: (1) recurrent neural network (RNN), (2) long short-term memory (LSTM) and (3) gated recurrent units (GRUs).

The details for each model are as follows:

8.3.1 SUPPORT VECTOR REGRESSION (SVR)

Support vector machines (SVMs) are a group of efficient supervised machine learning models capable of handling both linear and non-linear regression and classification tasks (Shah, 2020; Smola & Schölkopf, 2004). Based on the same principles as SVM, the SVR method is used for working with continuous-valued data instead of classification tasks (Vapnik et al., 1997). The intuitive approach for SVR is that it takes input data, and in order to process them via linear function, it applies mapping of input data to high-dimensional feature vector space through inbuilt non-linear function (Chuang et al., 2002). In the recent surge of forecasting models, SVR has stood its ground and has been successfully deployed in various fields with outstanding performance, becoming a standard out-of-the-box method in machine learning frameworks (Zhang et al., 2011).

Yan and Chowdhury (2013) presented the comparison between SVM and least squares support vector machine (LSSVM) to forecast Electricity-MCP (Market Clearing Price) data based on mid-term data. They also proposed another blended forecasting model for E-MCP by using SVM and the Autoregressive Moving Average with External input (ARMAX) (Yan & Chowdhury, 2013). From the above-mentioned applications, the applicability of SVR in different domains is evident. Finally, the major advantage of SVR over other forecasting methods is that it overcomes basic problems like over-fitting (because of high bias) and local minima (Yan & Chowdhury, 2014). Also, it is very simple to apply, unlike other models, which include complex modeling and structure designing. As already discussed above, the working principle of SVR (Cristianini & Shawe-Taylor, 2000) is to map the input feature x into a high-dimensional feature space via non-linear function φ as described below. Different from conventional statistical learning theory, SVR tries to find the best function f by minimizing the structural risk, and it also has moderate generation capability. ε-insensitive loss function used in the equation (8.3) calibrates the size of approximation error in order to control the number of support vectors and generation ability (Lin et al., 2007):

$$f(x) = w * \varphi(x) + b \tag{8.1}$$

$$\frac{1}{2}\|w\|^2 + C \sum_{i=1}^{l} \left(\xi_i + \xi_i^* \right) \tag{8.2}$$

$$s.\,t.\ ((w * \varphi(x)) + b) - y_i \leq \varepsilon + \xi_i, \quad i = 1,\ 2,\ ...,l \tag{8.3}$$

$$y_i - ((w * \varphi(x)) + b) \leq \varepsilon + \xi_i^*, \quad i = 1,\ 2,\ ...,l \tag{8.4}$$

$$\xi_i^{(*)} \geq 0,\ i = 1,\ 2,\ ...,\ l \tag{8.5}$$

where weights $w \in R^n$, bias $b \in R$ and ξ_i, ξ_i^* are called slack variables, $\xi_i^{(*)} \geq 0$ means $\xi_i^* \geq 0$ and $\xi_i \geq 0$, C is a pre-specified value (hyperparameter) that maintains balance between model complexity and error approximation.

8.3.2 Vector Auto Regression (VAR)

VAR is considered to be one of the most effective, pliable and easy-to-use practical models for time series forecasting. Around four decades ago, Christopher Sims (1980) provided this empirical macroeconomic framework, which demonstrated a new approach towards the statistical world. VAR is among those methods which utilize stochastic process models, a general extension of single-variable Autoregressive (AR) models to multivariate time series in order to develop linear interdependencies (Opgen-Rhein & Strimmer, 2007). Naturally, the model has been derived from the notion of Granger causality (Detto et al., 2012; Seghouane & Amari, 2012), i.e., a cause cannot come after results. It implies that the abundance of all data in time t_τ could be used to predict the abundance of data in time t, where τ is the time lag (Jiang et al., 2015). The applications of VAR are diverse, from economics to natural sciences.

The VAR is an n-equation, n-variable linear model which involves lagged value (the lagged values of a time series model are the values of the specified variables occurring prior to the present observation; Özcan & Öğüdücü, 2015) of its own variable plus the current and past values of other $(n - 1)$ variables. Amongst the set of all variables, often called endogenous variables, every variable of a model has its own equation, which tends to evolve over time in order to make better predictions. VARs provide a coherent and reliable approach to multivariate time series forecasting, data description, structural inference and policy analysis. These are easy to apply in practice, and their simple framework provides a schematic way to capture the rich dynamics of time series data.

Let n be the number of variables and $Y_t = (y_{1t}, y_{2t}, ..., y_{nt})'$ denote a time series vector of n dimensions. The basic p^{th} order vector autoregression model (Zivot & Wang, 2007) has the form:

$$Y_t = C + \prod_{y_{t-1}}^{1} + \prod_{y_{t-2}}^{2} + ... + \prod_{y_{t-p}}^{p} + \in_t; \ t = 1, \ ..., T \qquad (8.6)$$

where C is a n-dimensional vector of intercepts; Π^k, $k = 1, \ ..., p$ are $[n \times n]$ coefficient matrices; T is the maximum number of days considered; and $\in_t$ is a n-dimensional vector of errors with a white noise vector process which has unobservable zero mean, finite covariance, constant variance and no correlation with its past values. For evaluating model parameters and lag length of the pth order, the VAR model uses the Akaike Information Criterion (AIC).

8.3.3 Polynomial Regression (PR)

PR is a type of regression model for n^{th} degree polynomials. It gives a relationship between the dependent variable y and the independent variable x. It fits a wide range of curvatures. In cases where linear regression does not apply or the curve is non-linear, PR is used to fit and achieve minimum error. It can be applied for single or multiple variables. PR has a wide range of applications, with one of the popular applications being the automatic measurement of electromyography activity (Takada et al., 1995). Multivariate PR has its applications in data mining also (Sinha, 2013). The general equation for PR is given by:

$$Y = \theta_0 + \theta_1 * x + \theta_2 * x^2 + \theta_3 * x^3 + \theta_4 * x^4 + ... + \theta_n * x^n \qquad (8.7)$$

Here, n is the degree of the polynomial and θ_i is the weight for the ith of x. The equation can also be represented in the vector form with an additional error vector $\overrightarrow{\varepsilon}$ as:

$$\overrightarrow{y} = X * \overrightarrow{\theta} + \overrightarrow{\varepsilon} \qquad (8.8)$$

The PR model works like the Taylor series expansion of an unknown non-linear model. It is important that the order of PR be as low as possible. Various strategies are used to find the perfect order of polynomials like the forward selection procedure, backward elimination method, etc. The relationship between x and y variables is measured by a value R-squared. The R- squared values range from 0 to 1, where 0 means no-relation while 1 means complete correlation. The best fit line is decided by the degree of the polynomial. The PR model is usually affected by outliers, so outliers must be pre-processed before applying the algorithms. Another issue with PR is that of multicollinearity, and hence the model does not fit properly. The solution is to map variables to a higher-order space. Various methods like curvilinear component analysis, principal component analysis, etc., are used for dealing with multicollinearity. There is another problem of computing parameters in PR, which can be resolved by orthogonal polynomials.

8.3.4 RECURRENT NEURAL NETWORK (RNN)

RNNs are an integral part of various deep learning methods used to perceive correlations in time series forecasting. The hidden states of RNN are distributed through the model, which provides ease in predicting future events. It can make predictions effectively when the information passed to it through the input gate is in the same present state. The most important feature of RNN is that it can retain a bunch of information for a short time, which leads to a minor problem while training the model on time series data, yet provides a considerable increase in accuracy for some applications. One of the distinguishing features of RNNs is the architecture, which incorporates varying sequences of inputs and outputs, that is, the way connections are made internally with hidden layers to build recurrent connections (Salvaris et al., 2018). The complexity of the network is $O(N^2)$, so in order to work efficiently, the number of input neurons (N) must be considerably less than the number of hidden neurons. Collectively, the set of values of hidden neurons are conjoined inside state vector S in finite-dimensional space $[0, 1]^N$. A recurrent network accepts the time-ordered sequence of inputs and processes them dynamically through different layers of the network mentioned in (8.9), where g is the sigmoid discriminant function and W_{ijk} is the weights of the network, to make a considerable impact on the state vector S_j and input neurons I_k (Giles & Omlin, 1996).

$$S_j^{(t+1)} = g(\lambda_i); \quad \because \quad \lambda_i = \sum_{j,k} W_{ijk} * S_j^{(t)} * I_k^{(t)} \tag{8.9}$$

At each time step t, x_t and x_{t-1} serve as two inputs to RNN. Accordingly, the hidden neurons from every time step contribute equally to the continuous data input. This is because h_t is a function of $\{W_{xh} * (x_t + h_{t-1})\}$. The weights W_{xh} and W_{hh} can be considered equal in magnitude, that is, the assumption that the size of hidden neurons is the same as input neurons can be made because W_{xh} and W_{hh} share part of the parameters; technically, this situation tries to overcome the problem of overfitting with the help of parameter sharing and meanwhile provides a sufficient amount of freedom to any of the above inputs. Inputs in Figure 8.1 are by default considered to be partially dependent on one another, though this could easily mean that shared parameters easily comprehend parallelism through the inputs while non-shared parameters provide adequate degrees of freedom for remodeling (Diao et al., 2019).

As RNNs are only the basic variant of recurrent networks, they involve problems like vanishing gradients, which impede the training process of long data sequences. The gradients which carry the information keep on updating, and the parameter values keep on diminishing, which results in no further learning. Also, for the next state, there is a requirement for a hidden layer activation function of the previous state only, which is a serious disadvantage of RNNs. Due to the pitfalls, newer deep learning methods for time series forecasting like LSTM and GRU have replaced RNNs and resolved these issues. But, mainframes are built on the grounds of RNNs.

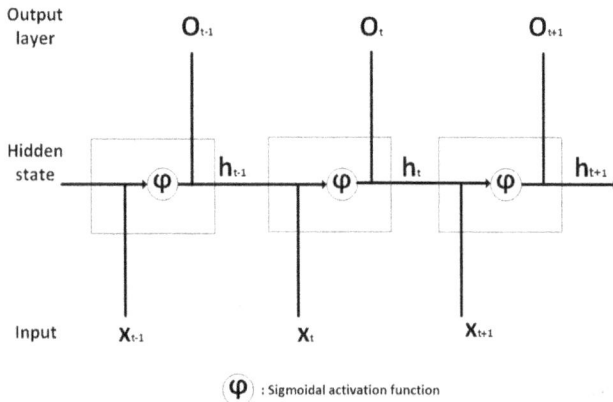

FIGURE 8.1 RNN structure for time series data at consecutive points of time.

8.3.5 LONG SHORT-TERM MEMORY (LSTM)

LSTM is a type of recurrent network used for sequence modeling. LSTM networks are good at handling short-term as well as long-term memories and hence are named as such. Unlike recurrent networks, LSTMs do not simply feed the outcome to memory but perform different mathematical operations and are capable of retaining long-term memory. Consequently, they have a wide range of applications like time series prediction and speech recognition. LSTM is also used in time series predictions, which involves long trends and random noise. RNN suffers from short memory and leaves information while dealing with big information, but LSTM preserves the information (Moghar & Hamiche, 2020).

As shown in Figure 8.2, LSTM has a chain-like structure with four interacting layers. It has four gates: forget, remember, learn and use. It has three inputs, namely short-term memory, long-term memory and training example (Li & Cao, 2018). When the inputs enter the LSTM model, they either go to the input gate or the learn gate. Here, the gate is a sigmoid neural network layer that controls the flow of information to or from memory. The gates are controlled by the output from the previous timestep and the current input. The forget gate controls the information to be thrown away from memory; the control gate controls the new information to be added; and the output/use gate decides the output from memory (Qiu et al., 2020).

The first step of LSTM is to decide what information has to be thrown away. It takes h_{t-1} and x_t as input and outputs 0 and 1 based on the decision whether the information is to be kept or thrown away. Further, in the next step, which information will be stored in the cell state is decided. This step is divided into two parts: the sigmoid layer decides the values to be updated while the tanh layer decides new values $\widetilde{C_t}$ to be added. These two steps are combined to update new states. The old C_{t-1} cell state is combined with C_t. For this, the forget gate f_t is

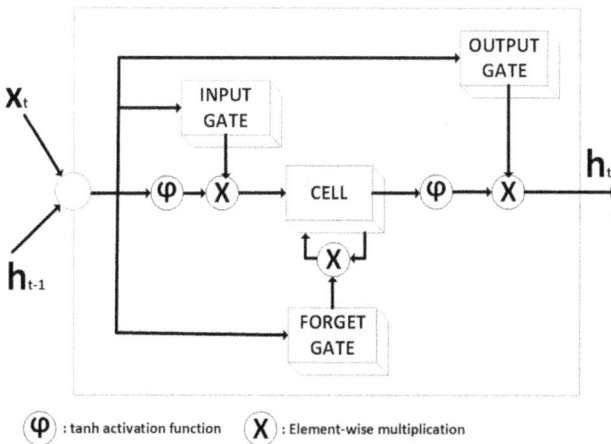

FIGURE 8.2   LSTM network structure with input, output and forget gates.

multiplied with the old state and added with the new candidate value $i_t \times \widetilde{C_t}$, which gives the value to be updated. Lastly, what values to output are decided, and a sigmoid layer is used to decide the part to be outputted. This output is passed through the *tanh* layer and multiplied with the sigmoid gate. The equations governing the LSTM cell are:

$$f_t = \sigma \left(W_f \cdot [h_{t-1}, x_t] + b_f \right) \qquad (8.10)$$

$$i_t = \sigma \left(W_i \cdot [h_{t-1}, x_t] + b_i \right) \qquad (8.11)$$

$$\widetilde{C_t} = tanh \left(W_c \cdot [h_{t-1}, x_t] + b_c \right) \qquad (8.12)$$

$$C_t = f_t \times C_{t-1} + i_t \times \widetilde{C_t} \qquad (8.13)$$

$$o_t = \sigma \left(W_o \cdot [h_{t-t}, x_t] + b_o \right) \qquad (8.14)$$

$$h_t = o_t \times tanh \left(C_t \right) \qquad (8.15)$$

8.3.6 GATED RECURRENT UNIT (GRU)

The GRU is a simple variation of the LSTM. In many cases, both generate comparable results with excellent accuracy. Similar to the LSTM units, the GRU cell includes gating blocks that control the flow of information inside the cell without the assistance of any separate memory units. Despite these similarities in the structure, which address the vanishing gradient problem faced in the vanilla RNN structure, there are differences in the structures of both the GRU and LSTM. As shown in Figure 8.3, the GRU cell has two gating layers—reset gate and update gate. The reset gate controls how much of the previous information from the network to forget. The update gate works analogously to the input and forget gate of an LSTM cell. It is responsible for the amount of information that passes from previous memory to the future.

Equations to determine the hidden state of a GRU cell at time step t, (h_t) are:

$$z_t = \sigma \left(W_t * X_t + U_z * h_{t-1} \right) \qquad (8.16)$$

$$r_t = \sigma \left(W_r * X_r + U_r * h_{t-1} \right) \qquad (8.17)$$

$$h_t' = tanh \left(W_h * X_t + r_t * U_h * h_{t-1} \right) \qquad (8.18)$$

$$h_t = (1 - z_t) * h_t' + z_t * h_{t-1} \qquad (8.19)$$

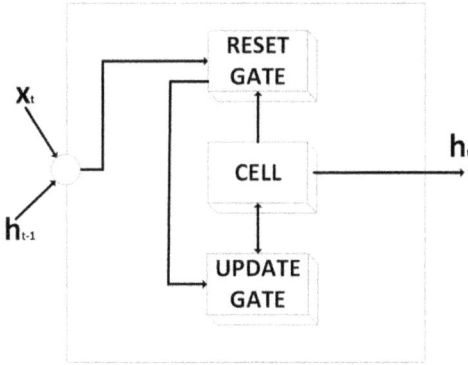

FIGURE 8.3   GRU cell with the interacting reset and update gates.

In the above equations, W and U represent the weight matrices of the network, σ is the sigmoid activation, z_t and r_t are the update and reset gate at time t, and h_t' is the hidden layer value. By observing equations (8.16)–(8.19) and Figure 8.3, it is evident that the number of parameters and gating classes are fewer in GRU when compared to LSTM or RNN. This facilitates easy training of the network and faster convergence.

8.4   DATA ANALYSIS

The raw data obtained from various sources is to be analyzed and pre-processed before feeding to the networks. This section elaborates on the variety of data and its conversion into useful data.

8.4.1   DATA DESCRIPTION

The data set used in this paper has been taken from covid-tracker source (COVID-19 India Org Data Operations Group, 2020). The data set is tallied with the official government data source and is highly reliable. However, the escalation and fall in cases depends upon many factors. For example, person-to-person spreads were subsequently increased due to recent freedom of movement in metropolitan cities. The line plots in Figure 8.4 are based on the daily figures of confirmed and re-covered cases in the entire Indian subcontinent, starting from the first reported case on February 3, 2020 to the tally as on October 20, 2020. For the training and evaluation of the forecasting models, the data has been split into training (from March 14, 2020 to October 20, 2020) and testing (from October 21, 2020 to November 4, 2020) sets. From Figure 8.4, it is evident that the virus spread ex-ponentially in the months leading up to September 2020, upon which it hit a certain peak and a steep downfall is visible. On November 4, 2020, the recorded cases were 8,363,329, recovered cases were 7,710,463 and deaths were 123,765. However, the number of cases is expected to increase in the coming days due to the complete

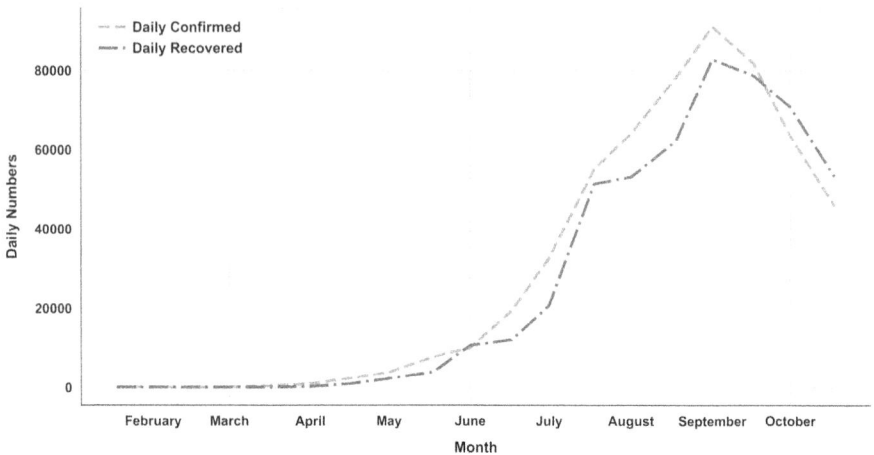

FIGURE 8.4   Daily confirmed and recovered case trend in India.

relaxation of lockdown norms and extensive inter-state movement due to the up-coming festival.

Figure 8.5 shows a comparative plot of the top ten worst-affected states by the COVID-19 pandemic. It includes the percentage of total confirmed, deceased and recovered cases in each state so as to give a broader view of the situation in India. Apparently, the four states— Maharashtra, Karnataka, Andhra Pradesh and Tamil Nadu—account for about 50% of the total reported and 60% of the total deceased cases in India. The total number of deaths in Maharashtra as of November 8, 2020 has reached 45,240, lying at the extreme end of all the affected states in India. On the other hand, its share of recoveries at 1,577,322 is proportionately lower. As shown in Figure 8.5, the significant difference in the percentage of deaths and recoveries hints at either a much more potent strain of the virus present in the state or a lack of adequate health care facilities. Kerala presents a contrasting scenario, with the deceased percentage of 2% having a considerable gap with the share of confirmed cases, owing to rapid testing and affordable facilities available in the state. Similarly, Odisha and Telangana have a similar situation to Kerala, with a comparatively low deceased ratio (2%). The raw data has been normalized by dividing each individual state case by the total cases in India for confirmed, deceased and recovered cases, respectively, to visualize and compare the statistics of COVID-19 in all the ten states. Visualization of the data in the unprocessed format is not suitable for analysis due to the widely varying numbers for each state; therefore, scaling of the data is necessary to obtain a proportionate comparison within the same range.

8.4.2   SEVERITY ANALYSIS

Figure 8.6 gives the severity analysis of the confirmed cases in all states and union territories of India. The geographical map is color-coded to show the comparative

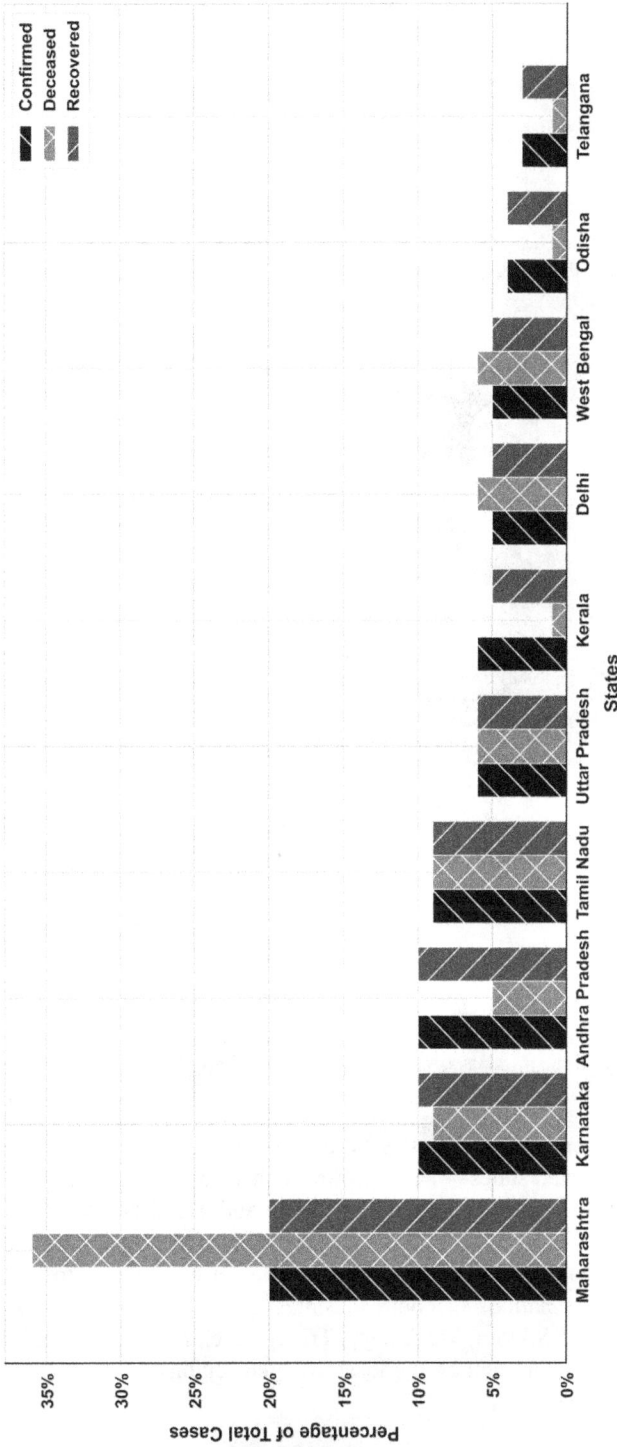

FIGURE 8.5 Comparative analysis of confirmed, deceased and recovered cases in the worst-affected Indian states (Maharashtra, Karnataka, Andhra Pradesh, Tamil Nadu, Uttar Pradesh, Kerala, Delhi, West Bengal, Odisha, Telangana).

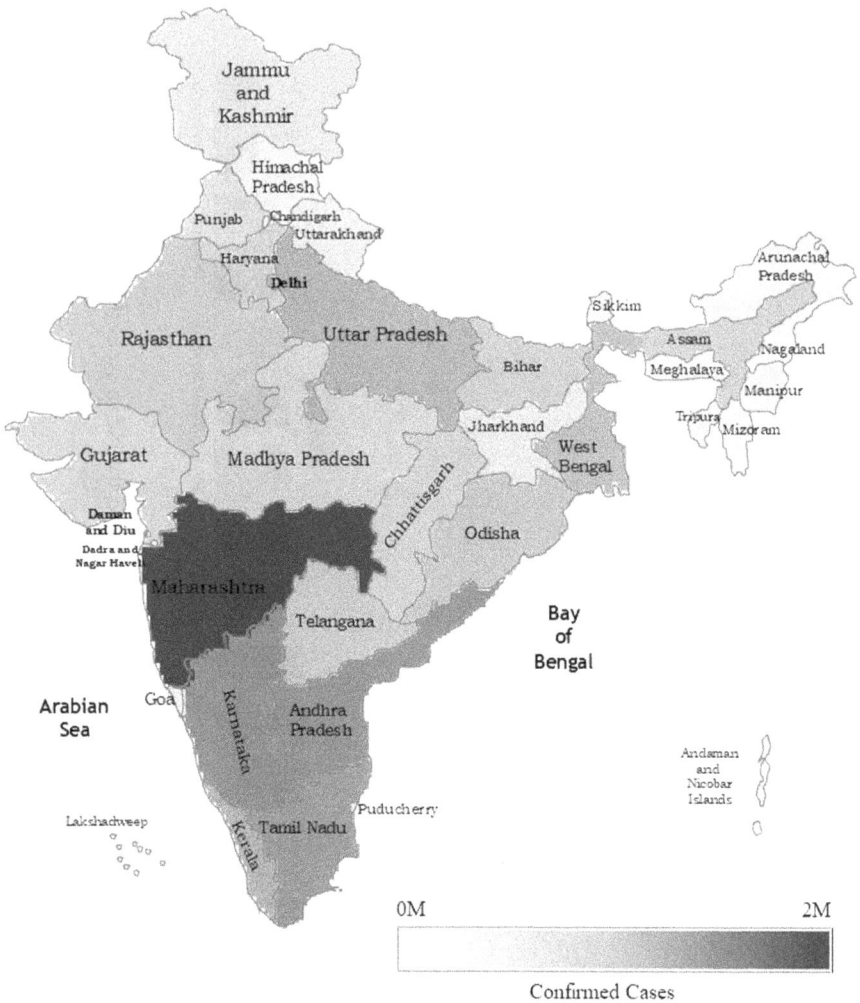

FIGURE 8.6 Classification of Indian states in different zones depending on the number of confirmed COVID-19 positive cases as of November 4, 2020.

distribution of cases from the most severely affected to the least affected state. The five worst-affected states— Maharashtra with 19.6%, Karnataka with 9.69%, Kerala with 5.99%, Andhra Pradesh with 9.60% and Tamil Nadu with 8.55% of total Indian cases—are on the highly affected zone of the spectrum of the heat map. These five states are together responsible for a majority of total cases reported in the entire nation. In contrast, states like Himachal Pradesh and the northeastern states of Sikkim, Meghalaya, Tripura, etc., have a negligible share of COVID-19 cases and are among the least-affected regions.

The country has been divided into three different regions based on the severity and frequency of the cases reported on a daily basis. The three zones are:

Zone 1: Highly affected region—High number of positive cases accompanied with a high doubling rate.
Zone 2: Mildly affected region—Comparatively fewer cases reported as compared to the red zone.
Zone 3: Lowly affected region—Almost zero cases reported in the last 21 days.

The classification of a region also depends on the factors such as doubling rate and the extent of testing in that particular region. The central as well as state governments have rolled out various restrictions for different zones in every unlock phase.

Figure 8.4 shows the comparison of daily confirmed cases reported to that of daily recovered cases of India. As observed, from mid-September, the recovered cases overtook reported cases, implying a decline in the number of active cases of the nation for the very first time since the inception of the first case in late January.

8.5 RESULTS AND DISCUSSIONS

The processed data is utilized for the implementation of the above-described methods for comparison and analysis. Each method is evaluated for metrics—MAE, RMLSE, MAPE and EV. The experimental set-up details and discussion are presented in this section.

8.5.1 EXPERIMENTAL SET-UP

For the experiments, we used the Jupyter notebook environment with Python version 3.7. The computational specifications include an Intel Core i5 eighth generation processor with 8 GB RAM and a 64-bit Windows 10 operating system. Since none of the experiments involved graphical processing units (GPUs) for training deep learning models, the details for the same have been omitted. The data processing and model design is done with the help of open-source libraries like Numpy (Oliphant, 2006), Pandas (McKinney, 2010), Scikit-Learn (Pedregosa et al., 2011) and PyTorch (Paszke et al., 2017). All the deep learning models (RNN, LSTM, GRU) are manually designed using functions and classes of the PyTorch framework, and the machine learning models (SVR, PR, VAR) are implemented using the Scikit-Learn and StatsModel libraries.

Except for PR, all the other forecasting models are trained on a univariate data set with the end goal of predicting daily confirmed cases for the next 15 days. All predictive models are trained and evaluated on the data of the ten states individually and their predictions, are compared based on the following performance metrics:

$$MAE = \frac{1}{n} * \sum_{k=1}^{n} \left| Y_k - \hat{Y}_k \right| \qquad (8.20)$$

$$RMSLE = \sqrt{\frac{1}{2}\sum_{k=1}^{n}\left(log\,Y_k - log\,\hat{Y}_k\right)^2} \tag{8.21}$$

$$MAPE = \frac{100}{n}*\sum_{k=1}^{n}\left|\frac{Y_k - \hat{Y}_k}{Y_k}\right| \tag{8.22}$$

$$EV = 1 - \frac{Var\,(Y - \hat{Y})}{Var\,(Y)} \tag{8.23}$$

where Y is the actual value, $\hat{Y}$ the predicted value and n denotes the total number of instances.

In the case of deep learning models, the optimal choice of optimization algorithms is key to avoid convergence to local minima and overshooting gradients. In this work the models are trained using the Adam (Kingma & Ba, 2014) and stochastic gradient descent (SGD) (Bottou, 2010) optimizers. The optimized models are then used to predict the number of confirmed cases for a period of 15 days (October 20 to November 4, 2020) and the performance of each model is logged using the output values of the four performance metrics described above. The best model for each state is selected based on the comparison of performance metrics.

8.5.2 DISCUSSIONS

This study includes six different forecasting methods to predict the confirmed COVID-19 cases in ten different states in India. Table 8.1. shares data from Maharashtra, Karnataka, Andhra Pradesh, Tamil Nadu, Uttar Pradesh, Kerala, Delhi, West Bengal, Odisha and Telangana and the performance of all the models on the above-mentioned metrics—MAE, RMLSE, MAPE and EV. Figures 8.7–8.16 present the actual data of confirmed cases, followed by the trend predicted by the best-performing model for each respective state.

As per the performances shown in Table 8.1, RNN outperforms every other model in the case of Maharashtra, Andhra Pradesh, West Bengal and Odisha, whereas PR performs better in the case of Karnataka, Tamil Nadu, Uttar Pradesh and Delhi. GRU was effective in predicting the cases of Kerala with an MAE of 1053.20, and VAR was successful in estimating cases of Telangana with an MAE of 246.63. It is worthwhile to consider here that the exceedingly good performance of PR is possibly due to the multivariate nature of data it is trained on. The performance of LSTM is moderate to worst (w r.t. all other models on all the metrics) in almost all states. The most decent performance of LSTM is in the case of Uttar Pradesh, accounting for an MAE error of 233.85 when compared to all the other states, and the worst performance of LSTM is for Maharashtra with an MAE of 2,148.75.

TABLE 8.1

Validation Metrics for Predicting COVID-19 Cases Using the SVR, VAR, PR, RNN, LSTM and GRU Models (The best performance for each state has been highlighted in bold.)

State	Model	Performance Metric			
		MAE	RMSLE	MAPE	EV
Maharashtra	SVR	1642.87	0.12	27.34	0.09
	VAR	1706.65	0.11	32.96	−0.53
	PR	985.67	0.07	19.36	−0.04
	RNN	**868.40**	**0.04**	**16.95**	**0.39**
	LSTM	2148.75	0.17	42.23	−2.21
	GRU	4562.67	0.40	84.50	−0.13
Karnataka	SVR	767.93	0.07	18.17	0.69
	VAR	1007.75	0.09	22.77	−0.09
	PR	**635.07**	**0.03**	**16.37**	**0.67**
	RNN	941.60	0.07	25.87	0.73
	LSTM	1780.36	0.20	51.38	-0.36
	GRU	701.53	0.05	18.21	0.44
Andhra Pradesh	SVR	650.13	0.10	21.66	0.01
	VAR	821.98	0.11	26.25	0.11
	PR	436.33	0.05	16.83	−0.10
	RNN	**387.13**	**0.04**	**15.46**	**0.20**
	LSTM	1056.28	0.16	41.98	−0.99
	GRU	2380.40	2.55	79.16	0.03
Tamil Nadu	SVR	289.20	0.02	10.61	0.72
	VAR	305.78	0.02	10.45	−0.73
	PR	**97.60**	**0.00**	**3.63**	**0.76**
	RNN	133.67	0.00	4.81	0.63
	LSTM	1274.01	0.16	47.71	−0.42
	GRU	502.20	0.03	18.44	0.25
Uttar Pradesh	SVR	332.87	0.04	16.08	0.08
	VAR	221.85	0.02	10.61	−0.40
	PR	**163.47**	**0.01**	**8.46**	**0.11**
	RNN	171.33	0.01	8.41	0.15
	LSTM	233.85	0.03	11.40	−0.11
	GRU	692.27	0.10	35.56	−0.24
Kerala	SVR	2039.67	0.18	34.06	−2.58
	VAR	1911.04	0.13	34.38	−0.27
	PR	1677.73	0.13	27.74	−1.63
	RNN	1128.47	0.06	20.23	−0.09
	LSTM	1415.45	0.09	24.79	−0.65
	GRU	**1053.20**	**0.04**	**16.07**	**0.08**

(Continued)

TABLE 8.1 (Continued)

State	Model	Performance Metric			
Delhi	SVR	853.73	0.06	18.08	0.17
	VAR	1451.95	0.12	35.57	−1.13
	PR	**782.07**	**0.05**	**17.41**	**0.19**
	RNN	975.40	0.07	20.09	0.28
	LSTM	1735.10	0.29	33.22	−0.68
	GRU	899.13	0.05	18.69	0.09
West Bengal	SVR	219.80	0.00	5.41	0.20
	VAR	119.77	0.00	2.96	−1.54
	PR	63.60	0.00	1.57	−0.07
	RNN	**57.33**	**0.00**	**1.42**	**0.08**
	LSTM	1029.50	0.14	25.60	−53.98
	GRU	598.60	0.02	14.87	0.01
Odisha	SVR	245.13	0.05	14.93	−0.09
	VAR	259.48	0.04	15.40	−0.20
	PR	174.87	0.02	11.56	0.04
	RNN	**164.00**	**0.02**	**11.05**	**0.54**
	LSTM	462.16	0.10	31.82	−1.81
	GRU	270.13	0.04	17.18	−1.15
Telangana	SVR	362.47	0.22	36.36	−1.48
	VAR	**246.63**	**0.08**	**24.29**	**0.07**
	PR	298.93	0.15	31.56	−0.94
	RNN	252.60	0.09	26.08	0.01
	LSTM	387.25	0.20	38.21	−1.56
	GRU	365.93	0.32	32.03	−1.32

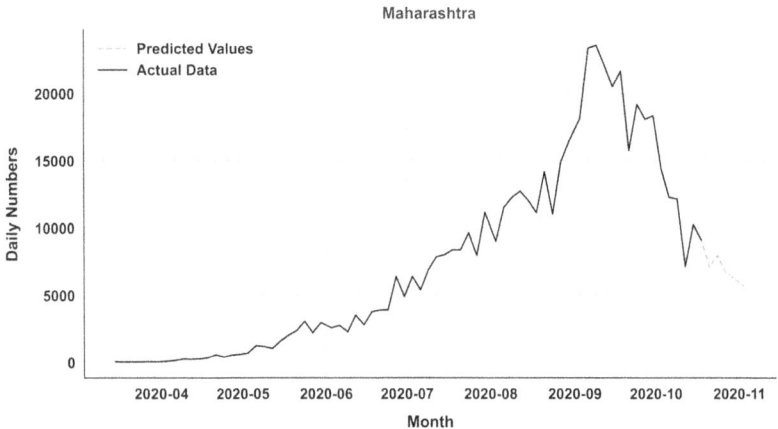

FIGURE 8.7 Time series forecasting for Maharashtra using RNN.

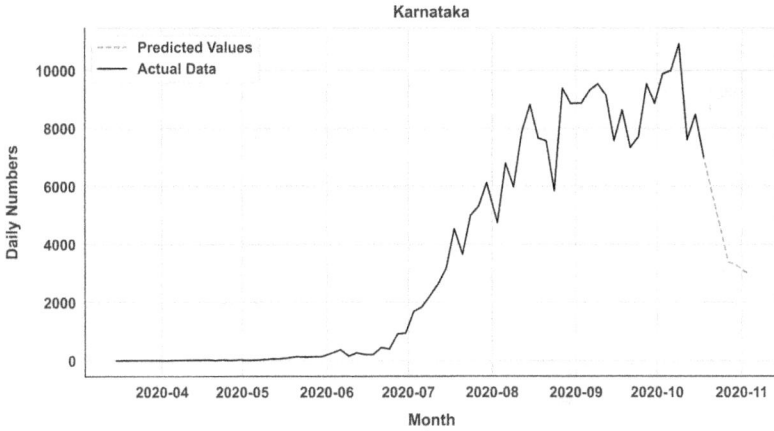

FIGURE 8.8   Time series forecasting for Karnataka using PR.

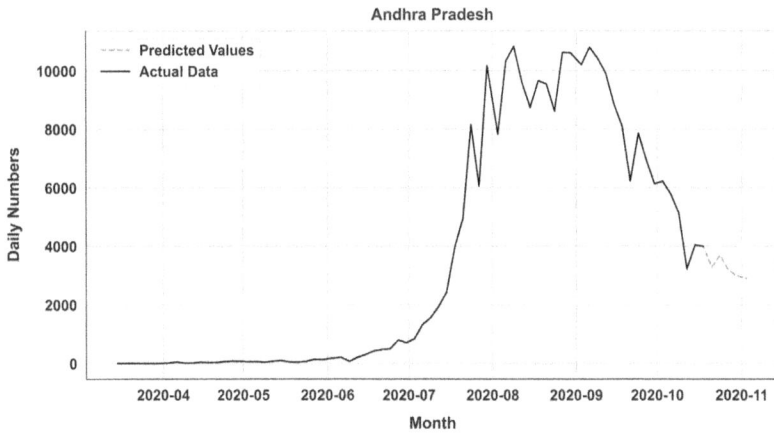

FIGURE 8.9   Time series forecasting for Andhra Pradesh using RNN.

When MAE values of Maharashtra state are compared for all the models, the performance of SVR and VAR models are very much comparable. GRU performs worst in the case of Maharashtra, with a huge margin on every performance metric (e.g., MAE is 4562.67). RNN performs best in the case of West Bengal, with an error of 57.33 on the MAE metric and a stunning 0.00 on the RMLSE metric. As evident from Table 8.1, the performance of SVR and VAR models is very close in almost every state. The MAE metric gives a detailed overview of the model performance for each case because of a huge margin that the RMLSE fails to provide.

Observing the prediction curves for Maharashtra (Figure 8.7), Andhra Pradesh (Figure 8.9), West Bengal (Figure 8.14) and Odisha (Figure 8.15), the RNN model

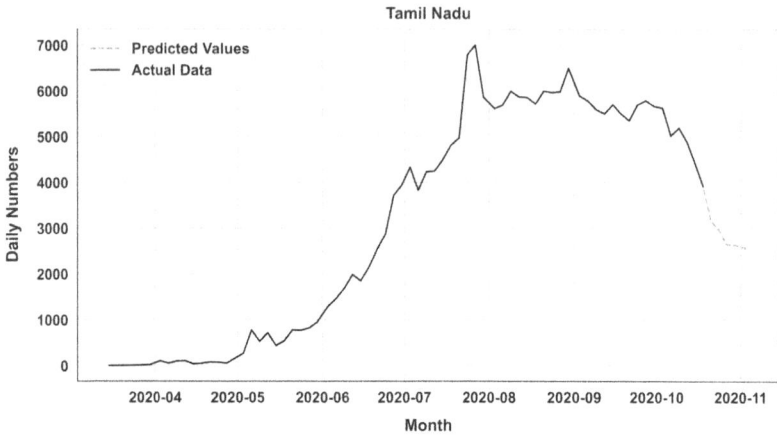

FIGURE 8.10   Time series forecasting for Tamil Nadu using PR

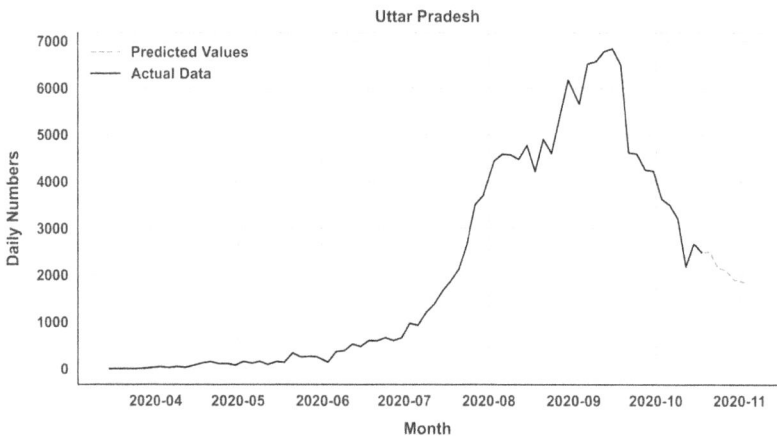

FIGURE 8.11   Time series forecasting for Uttar Pradesh using PR

seems to perform better in states which display a slightly linear behavior in their initial period and also in their later period, having achieved their peak already and now facing a downward trend. As seen in the prediction curve for West Bengal (Figure 8.14), it is a clear exception to this statement where only a continuous upward yet less-vigorous trend is visible and no peak is observed, yet RNN seems to perform better. In states where plateau-like curves are observed, PR outperforms its counterparts in the states of Karnataka (Figure 8.8), Tamil Nadu (Figure 8.10), Uttar Pradesh (Figure 8.11) and Delhi (Figure 8.13). PR is able to account for the drastic variations in the data of Delhi (Figure 8.13) and Karnataka (Figure 8.8), where the frequency of daily confirmed cases is rapidly changing with major spikes

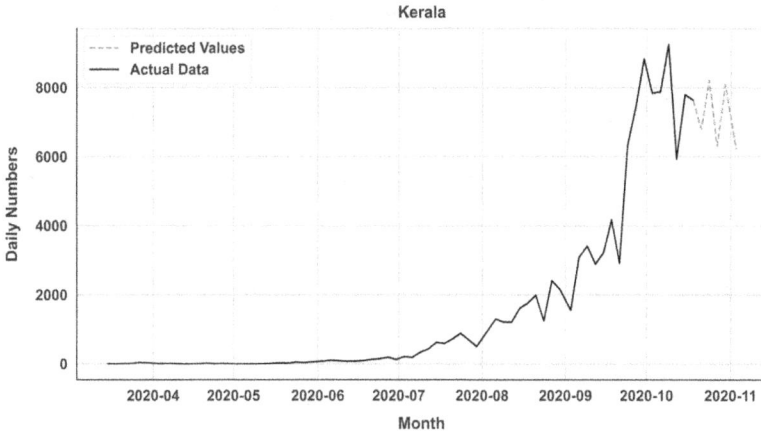

FIGURE 8.12 Time series forecasting for Kerala using GRU.

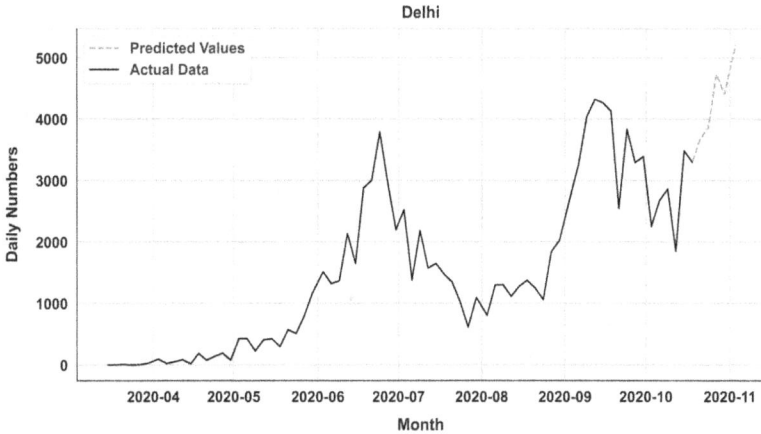

FIGURE 8.13 Time series forecasting for Delhi using PR.

and dips; Delhi already displaying two major peaks, the first peak in the month of June–July with approximately 4,000 confirmed cases per day and the second peak in the month of September–October, with about 4,100 confirmed cases per day. As per the predicted trend, the cases are bound to increase in the upcoming days with a third peak in the month of November or December. After successfully tackling the pandemic in the early days, Kerala (Figure 8.12) is currently facing an upsurge in the daily confirmed cases with 6,000–8,500 daily new cases in the month of October, with the predicted confirmed daily cases in the range of 6,000–8,000 cases per day for the month of November.

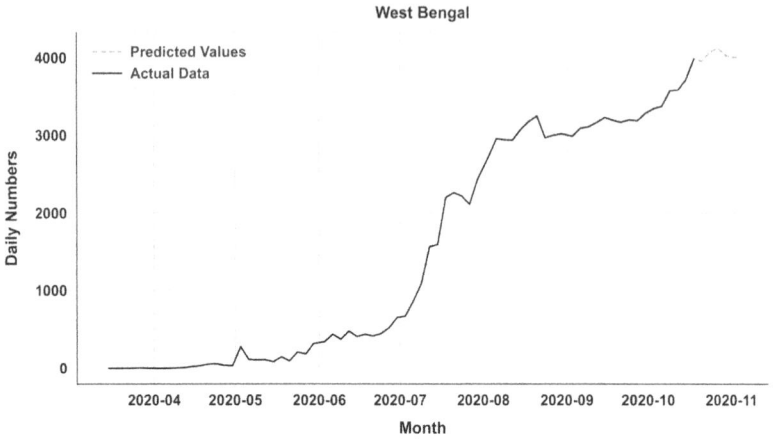

FIGURE 8.14 Time series forecasting for West Bengal using RNN.

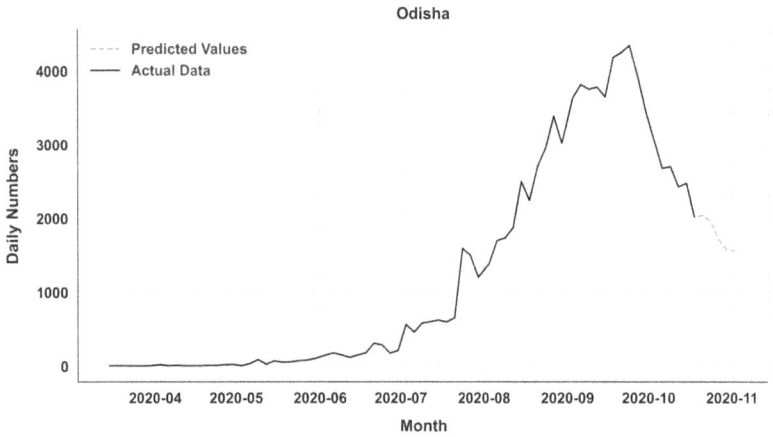

FIGURE 8.15 Time series forecasting for Odisha using RNN.

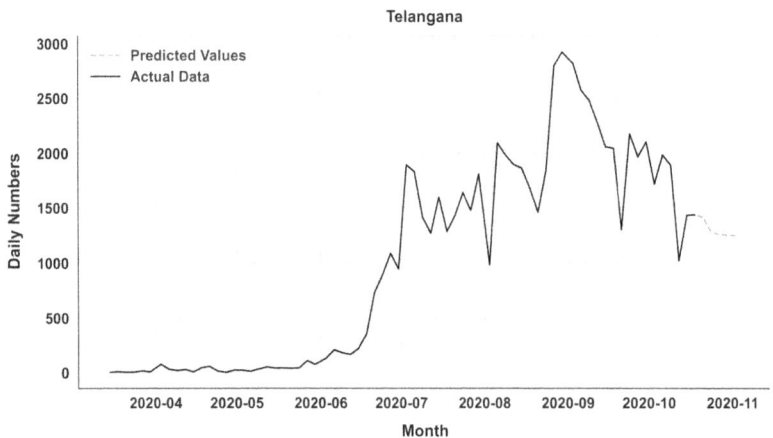

FIGURE 8.16 Time series forecasting for Telangana using VAR.

8.6 CONCLUSIONS

This chapter presents a comparative study of six different time series forecasting methods for predicting the number of COVID-19 positive cases in Indian states. It provides a detailed analysis of the current situation with graphical plots to better understand the situation. Based on the reported cases and rate of spread, Indian states and union territories have been classified into different zones for effective management. Focusing on three machine learning models and three deep learning models, we provided a detailed mathematical background for all referred methods. The research is carried out solely on the basis of statistical data and modeling, the influence of the preventive steps taken during the outbreak and the conformity with the regulations on sanitation has been neglected. Nevertheless, the methods used are widely regarded and reliable and offer a realistic prediction for the near future.

REFERENCES

Arora, P., Kumar, H., & Panigrahi, B. K. (2020). Prediction and analysis of COVID-19 positive cases using deep learning models: A descriptive case study of India. *Chaos, Solitons & Fractals, 139*, 110017.

Bottou, L. (2010). Large-scale machine learning with stochastic gradient descent. In *Proceedings of COMPSTAT'2010* (pp. 177–186). Physica-Verlag HD.

Chadsuthi, S., Modchang, C., Lenbury, Y., Iamsirithaworn, S., & Triampo, W. (2012). Modeling seasonal leptospirosis transmission and its association with rainfall and temperature in Thailand using time–series and ARIMAX analyses. *Asian Pacific Journal of Tropical Medicine, 5*(7), 539–546.

COVIDIndia.org. (2020). COVID-19 Tracker Updates for India. Retrieved November 04, 2020 from https://covidindia.org/.

Chimmula, V. K. R., & Zhang, L. (2020). Time series forecasting of COVID-19 transmission in Canada using LSTM networks. Chaos. *Solitons & Fractals*, 135, 109864. doi:10.101 6/j.chaos.2020.109864

Chuang, C. C., Su, S. F., Jeng, J. T., & Hsiao, C. C. (2002). Robust support vector regression networks for function approximation with outliers. *IEEE Transactions on Neural Networks, 13*(6), 1322–1330.

COVID-19 India Org Data Operations Group. 2020. covid19india org 2020 tracker. Retrieved November 08, 2020, from https://api.covid19india.org/.

Craven, J. (2020, November 03). COVID-19 vaccine tracker. Retrieved November 03, 2020, from https://www.raps.org/news-and-articles/news-articles/2020/3/covid-19-vaccine-tracker.

Cristianini, N., & Shawe-Taylor, J. (2000). *An introduction to support vector machines and other kernel-based learning methods*. Cambridge University Press.

Detto, M., Molini, A., Katul, G., Stoy, P., Palmroth, S., & Baldocchi, D. (2012). Causality and persistence in ecological systems: a nonparametric spectral Granger causality approach. *The American Naturalist, 179*(4), 524–535.

Diao, E., Ding, J., & Tarokh, V. (2019, December). Restricted recurrent neural networks. In *2019 IEEE International Conference on Big Data (Big Data)* (pp. 56–63). IEEE.

Economic Times (2020). *Covid-19 recovery rate*. Retrieved November 04, 2020, from https://economictimes.indiatimes.com/news/politics-and-nation/covid-19-recovery-rate-at-41–61-as-compared-to-11-4-on-april-15-govt/articleshow/76016083.cms

Fanelli, D., & Piazza, F. (2020). Analysis and forecast of COVID-19 spreading in China, Italy and France. *Chaos, Solitons & Fractals, 134*, 109761.

Geneva: World Health Organization (2020). WHO coronavirus disease (COVID-19) dashboard. Retrieved November 4, 2020, from https://covid19.who.int/.

Giles, C. L., & Omlin, C. W. (1996). Rule revision with recurrent networks. *IEEE Transactions on Knowledge and Data Engineering*, 8(1), 183–197.

Golechha, M. (2020). COVID-19 containment in Asia's largest urban slum Dharavi-Mumbai, India: Lessons for policymakers globally. *Journal of Urban Health*, 97(6), 796–801.

Hanf, M., Adenis, A., Nacher, M., & Carme, B. (2011). The role of El Nino Southern Oscillation (ENSO) on variations of monthly Plasmodium falciparum malaria cases at the Cayenne General Hospital, 1996–2009, French Guiana. *Malaria Journal*, 10(1), 100.

Jia, L., Li, K., Jiang, Y., & Guo, X. (2020). Prediction and analysis of Coronavirus Disease 2019. arXiv preprint arXiv: 2003.05447.

Jiang, X., Hu, X., & He, T. (2015, November). Time series analysis of microbiome data regularized by local linear manifolds. In *2015 IEEE International Conference on Bioinformatics and Biomedicine (BIBM)* (pp. 119–122). IEEE.

Kingma, D. P., & Ba, J. (2014). Adam: A method for stochastic optimization. arXiv preprint arXiv:1412.6980.

Klompas, M. (2020). Coronavirus disease 2019 (COVID-19): Protecting hospitals from the invisible.

Kucharski, A. J., Russell, T. W., Diamond, C., Liu, Y., Edmunds, J., Funk, S. & Eggo, R. (2020). Early dynamics of transmission and control of COVID-19: A mathematical modelling study. *The Lancet Infectious Diseases*, 20(5), 553–558.

Li, Y., & Cao, H. (2018). Prediction for tourism flow based on LSTM neural network. *Procedia Computer Science*, 129, 277–283.

Lin, K., Lin, Q., Zhou, C., & Yao, J. (2007, August). Time series prediction based on linear regression and SVR. In *Third International Conference on Natural Computation (ICNC 2007)* (Vol. 1, pp. 688–691). IEEE.

McKinney, W. (2010, June). Data structures for statistical computing in python. In Proceedings of the 9th Python in Science Conference (Vol. 445, pp. 51–56).

Moghar, A., & Hamiche, M. (2020). Stock market prediction using LSTM recurrent neural network. *Procedia Computer Science*, 170, 1168–1173.

Oliphant, T. E. (2006). *A guide to NumPy* (Vol. 1, p. 85). Trelgol Publishing.

Opgen-Rhein, R., & Strimmer, K. (2007). Learning causal networks from systems biology time course data: an effective model selection procedure for the vector autoregressive process. *BMC Bioinformatics*, 8(2), 1–8.

Özcan, A., & Öğüdücü, Ş. G. (2015, June). Multivariate temporal link prediction in evolving social networks. In *2015 IEEE/ACIS 14th International Conference on Computer and Information Science (ICIS)* (pp. 185–190). IEEE.

Papastefanopoulos, V., Linardatos, P., & Kotsiantis, S. (2020). COVID-19: A comparison of time series methods to forecast percentage of active cases per population. *Applied Sciences*, 10(11), 3880.

Paszke, A., Gross, S., Chintala, S., Chanan, G., Yang, E., DeVito, Z., Lin, Z., Desmaison, A., Antiga, L., & Lerer, A. (2017). Automatic differentiation in pytorch.

Pedregosa, F., Varoquaux, G., Gramfort, A., Michel, V., Thirion, B., Grisel, O., Blondel, M., Prettenhofer, P., Weiss, R., Dubourg, V., Vanderplas, J., Passos, A., Cournapeau, D., Brucher, M., Perrot, M., & Duchesnay, É. (2011). Scikit-learn: Machine learning in Python. *The Journal of Machine Learning Research*, 12, 2825–2830.

Preiser, W., Van Zyl, G., & Dramowski, A. (2020). COVID-19: Getting ahead of the epidemic curve by early implementation of social distancing. *SAMJ: South African Medical Journal*, 110(4), 258.

Qiu, J., Wang, B., & Zhou, C. (2020). Forecasting stock prices with long-short term memory neural networks based on attention mechanism. *PloS One*, 15(1), e0227222.

Ray, A. (2020). World's 'first' COVID-19 vaccine Sputnik V out in Russia: How it works and who will get it? Retrieved November 08, 2020, from https://www.livemint.com/news/india/world-s-first-covid-19-vaccine-out-in-russia-how-it-works-and-who-will-get-it-11597143985840.html.

Rustam, F., Reshi, A. A., Mehmood, A., Ullah, S., On, B., Aslam, W., & Choi, G. S. (2020). COVID-19 future forecasting using supervised machine learning models. *IEEE Access*, 8, 101489–101499.

Saba, A. I., & Elsheikh, A. H. (2020). Forecasting the prevalence of COVID-19 outbreak in Egypt using non-linear autoregressive artificial neural networks. *Process Safety and Environmental Protection*, 141, 1–8.

Salvaris, M., Dean, D., & Tok, W. H. (2018). Recurrent Neural Networks. In *Deep Learning with Azure* (pp. 161–186). Apress.

Seghouane, A. K., & Amari, S. I. (2012). Identification of directed influence: Granger causality, Kullback-Leibler divergence and complexity. *Neural Computation*, 24(7), 1722–1739.

Shah, R. (2020, June). Solar cell parameters extraction using multi-target regression methods. In *2020 IEEE International Conference on Environment and Electrical Engineering and 2020 IEEE Industrial and Commercial Power Systems Europe (EEEIC/I&CPS Europe)* (pp. 1–6). IEEE.

Sinha, P. (2013). Multivariate polynomial regression in data mining: methodology, problems and solutions. *International Journal of Scientific and Engineering Research*, 4(12), 962–965.

Smola, A. J. (1998). B. Sch olkopf. A tutorial on support vector regression. *Statistics and Computing*.

Smola, A. J., & Schölkopf, B. (2004). A tutorial on support vector regression. *Statistics and Computing*, 14(3), 199–222.

Song, X., Xiao, J., Deng, J., Kang, Q., Zhang, Y., & Xu, J. (2016). Time series analysis of influenza incidence in Chinese provinces from 2004 to 2011. *Medicine*, 95(26), 3929–3936.

Takada, K., Yashiro, K., & Morimoto, T. (1995). Application of polynomial regression modeling to automatic measurement of periods of EMG activity. *Journal of Neuroscience Methods*, 56(1), 43–47.

Times of India (2020). India's Covid-19 Tally. Retrieved November 04, 2020, from https://timesofindia.indiatimes.com/india/indias-covid-19-tally-races-past-83-lakh-national-recovery-rate-reaches-92-09/articleshow/79034528.cms

Toğaçar, M., Ergen, B., & Cömert, Z. (2020). COVID-19 detection using deep learning models to exploit Social Mimic Optimization and structured chest X-ray images using fuzzy color and stacking approaches. *Computers in Biology and Medicine*, 121, 103805.

Tsay, R. S. (2005). *Analysis of financial time series* (Vol. 543). John Wiley & Sons.

Tuli, S., Tuli, S., Tuli, R., & Gill, S. S. (2020). Predicting the growth and trend of COVID-19 pandemic using machine learning and cloud computing. *Internet of Things*, 11, 100222.

Vapnik, V., Golowich, S. E., & Smola, A. J. (1997). Support vector method for function approximation, regression estimation and signal processing. In *Advances in neural information processing systems* (pp. 281–287).

Wang, L., Li, J., Guo, S., Xie, N., Yao, L., Cao, Y., Day, S., Howard, S., Graff, J., Gu, T., Ji, J., Gu, W. & Sun, D. (2020). Real-time estimation and prediction of mortality caused by COVID-19 with a patient information based algorithm. *Science of the Total Environment*, 727, 138394.

WHO India. 2020. Responding to COVID-19 – Learnings from Kerala. Retrieved November 08, 2020 from https://www.who.int/india/news/feature-stories/detail/responding-to-covid-19---learnings-from-kerala.

Wikipedia contributors (2020, November 18). Coronavirus disease 2019. In *Wikipedia, The Free Encyclopedia*. Retrieved 09:03, November 18, 2020, from https://en.wikipedia.org/w/index.php?title=Coronavirus_disease_2019&oldid=989303232.

Wikipedia contributors (2020, November 16). COVID-19 pandemic in India. In *Wikipedia, The Free Encyclopedia*. Retrieved 09:04, November 18, 2020, from https://en.wikipedia.org/w/index.php?title=COVID-19_pandemic_in_India&oldid=988941142.

World Health Organization (2020). COVID-19 and food safety: guidance for food businesses. Interim guidance, 7 April 2020.

Wu, J. T., Leung, K., & Leung, G. M. (2020). Nowcasting and forecasting the potential domestic and international spread of the 2019-nCoV outbreak originating in Wuhan, China: A modelling study. *The Lancet*, *395*(10225), 689–697.

Yan, X., & Chowdhury, N. A. (2013). A comparison between SVM and LSSVM in mid-term electricity market clearing price forecasting. 2013 26th IEEE Canadian Conference on Electrical and Computer Engineering (CCECE). 10.1109/ccece.2013.6567685.

Yan, X., & Chowdhury, N. A. (2014). Mid-term electricity market clearing price forecasting utilizing hybrid support vector machine and auto-regressive moving average with external input. *International Journal of Electrical Power & Energy Systems*, *63*, 64–70. https://doi.org/10.1016/j.ijepes.2014.05.037.

Yin, R., Luusua, E., Dabrowski, J., Zhang, Y., & Kwoh, C. K. (2020). Tempel: time-series mutation prediction of influenza A viruses via attention-based recurrent neural networks. *Bioinformatics*, *36*(9), 2697–2704.

Zeroual, A., Harrou, F., Dairi, A., & Sun, Y. (2020). Deep learning methods for forecasting COVID-19 time-series data: A comparative study. *Chaos, Solitons & Fractals*, *140*, 110121.

Zhang, Y., Tian, W., & Liu, S. (2011, May). Fire time series forecasting based on Markov-SVR model. In *2011 International Conference on Multimedia and Signal Processing* (Vol. 2, pp. 278–281). IEEE.

Zivot, E., & Wang, J. (2007). *Modeling financial time series with S-Plus®* (Vol. 191). Springer Science & Business Media.

9 Time Series Forecasting Using Support Vector Machines

V. Kumar[1] and S. Yadav[2]
[1]Department of Mathematics, Aksum University, Aksum, Ethiopia,
[2]NorthCap University, Gurugram Haryana-122017, India

9.1 INTRODUCTION

The focus of this chapter is forecasting the outcome values of time series using a support vector machine (SVM). Many researchers are performing active research on time series forecasting, and this makes time series modeling a prime tool for prediction (Cochrane, 1997). Therefore, several mathematical models have been proposed to decrease inaccuracy and increase efficiency of time series modeling and forecasting (Cottrell et al., 1995). There are many pros and cons for all these methods used for forecasting, but in view of their usability, we cannot ignore their future importance in the real world (Gareth & Swift, 1993; Jenkins et al., 2016).

Currently support vector machine (SVM) is widely used by a plethora of researchers and industrial experts for future estimation. This technique was first introduced to find the classified solution of prototype text arrangement, face detection and optimal character identification by Vapnik et al. (1996) at the AT&T Bell Test Center. Nevertheless, shortly thereafter they initiated extensive applications in other areas. SVM works as a supervised machine learning algorithm established in statistical learning theory and is accepted in statistics and signal processing. SVM is also effective in high-dimensional spaces, even if the number of samples is less than the dimensions (Cortes & Vapnik, 1995).

An important alternative property of SVM is that its solutions are always globally optimal and unique because the training process to apply this method is almost as similar and easy to understand as solving a linearly controlled quadratic programming problem. However, despite so many advantages, SVM has a disadvantage. When the sample size is large, the method may be inaccurate, and the time complexity of the solution will increase due to the need for more computation (Vapnik, 1998).

In this chapter, we explain the important concepts and basics requirements of SVM and apply it to real-life models.

DOI: 10.1201/9781003102281-9

9.2   INTRODUCTION TO STATISTICAL LEARNING THEORY

The objectives of statistical learning are understanding and forecasting. Learning falls into numerous classes, including directed learning, solo learning, web-based learning and fortification learning. From the point of view of measurable learning hypothesis, administered learning is the best. Supervised taking-in includes learning a prepared set of facts. Each point in the set is an info yield pair, where the information leads to a yield. The learning issue involves surmising the capacity that maps between the information and the yield, with the end goal that the education capacity can be utilized to foresee the yield from future information (Vapnik, 1998).

Statistical learning hypothesis, given by Vapnik, is used to determine a learning strategy (Vapnik & Chervonenkis, 1971). As indicated by Vapnik et al. (1996), there are three primary issues in AI: characterization, regression and density estimation. As per the consideration of such cases, the information is prepared utilizing the learning machine, and afterward general outcomes are surmised dependent on the information. Now, we choose a possible input vector space X, and output vector space would be Y. Then, according to statistical learning theory (Campbell and Walker 1977), in the product space there is some unknown probability distribution. $P(z) = p(\alpha, \beta)$ is some unknown for $Z = X \times Y$ for this probability distribution; we have a training set of n samples denoted as:

$$Z = \left\{ (\alpha_1, \beta_1),\ (\alpha_2, \beta_2),\ ... (\alpha_n, \beta_n) \right\} = \{z_1, z_2 ... z_n\} \qquad (9.1)$$

This input and output concept is also showed in the diagram in Figure 9.1.

9.3   SUPPORT VECTOR MACHINE (SVM)

We can say that SVM classifiers are best used to differentiate between two lines or hyper-planes. The main task is to identify the best or right hyper-plane. For the SVM classifier, finding a linear hyper-plane between two classes is not difficult (Parrelli, 2001; Suykens & Vandewalle, 2000). The kernel technique trick would help us to choose the best hyper-plane using an algorithm. These kernel functions

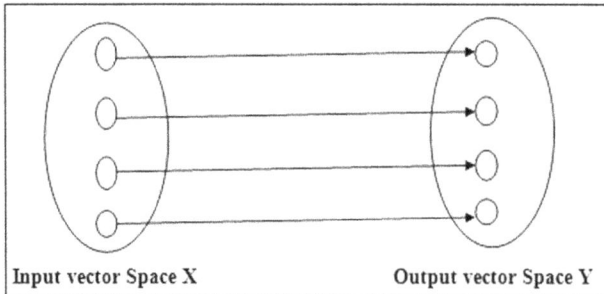

FIGURE 9.1   Probabilistic mapping of input and output points.

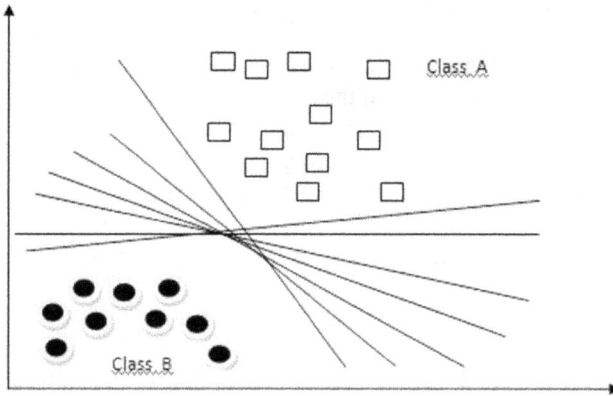

FIGURE 9.2 Infinite number of linearly separating hyper-planes.

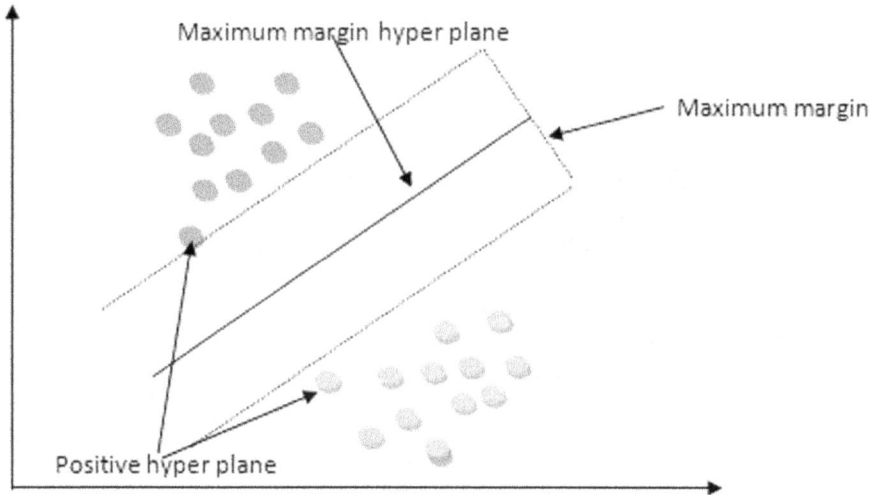

FIGURE 9.3 The maximum margin hyper-plane.

are being used as a transformation from a low dimensional input space to another higher dimensional output space (Mueller et al., 1995; Mukherjee et al., 1995). This is a very useful task as the output would be a separable problem from a not separable problem, shown in Figures 9.2 and 9.3.

In simple words, SVM transforms extremely complex data and helps to separate the data based on labels or defined outputs. In Figure 9.2, we consider two different linearly separable data points in n-dimensional vector space. These data sets are classified into two different classes, A and B, using multiple hyper-planes; we presented a diagrammatic view for the statement. In Figure 9.3, we have a clear view that both the classes are separated by the maximum margin hyper-plane with a margin distance.

9.4 EMPIRICAL RISK MINIMIZATION (ERM)

There are many uses for the statistical learning theory, but the main aim is to find a suitable estimator function φ from the input space X to the output space Ω such that it associates each point of X to a point on Ω. Vapnik and Chervonenkis (1971), gave a theory of statistical learning to find the common error. They used $X \times \Omega$ to find the real and expected output common errors. This common error comes as we are using an estimator function φ. Hence, out of these functions, we need to choose a most suitable function which minimizes this risk (Mueller et al., 1995).

Now, consider input space $X \subseteq R^n$, and for the output space $\Omega \subseteq R$, choose the set of functions $F = \varphi(\alpha, \omega)$, where ω is the parameters which define φ. The function F defines on input space X, which projects the points into the output space Y. For an input vector α, assume that β is an output vector. Let us define the estimated risk $R(\varphi)$ as

$$R(\phi) = \int L(\beta, \varphi(\alpha, \omega)) dP(\alpha, \beta) \tag{9.2}$$

The R(φ) is function of prediction function φ, which is defined as integral of $L(\beta, \varphi(\alpha, \omega))$. Where $L(\beta, \varphi(\alpha, \omega))$ is corresponding error between the actual value β and the predicted value $\varphi(\alpha, \omega)$ (Raicharoen & Lursinsap, 2002). There are many different ways to define loss function $P(\alpha, \beta)$ (Adhikari & Agrawal, 2013). Now to minimize the estimated risk $R(\varphi)$ define the target function φ_0, which is actually the most appropriate prediction function.

Hence, we need to find the target function, which is the deal estimator. As the probability distribution function $P(\alpha, \beta)$ is not defined, it is very difficult to calculate the actual risk. The Empirical Risk Minimization (ERM) principle has been suggested by Vapnik to overcome the above problem (Cortes & Vapnik, 1995).

The concept of ERM is used to calculate the estimated risk $R(\varphi)$. The value of $R(\varphi)$ is calculated by using the training set. This approximate value of the estimated risk $R(\varphi)$ is known as empirical risk. Consider any given value of the training set $\{\alpha_i, \beta_i\}$ where, $\alpha_i \in X \subseteq R^n$ and $\beta_i \in \Omega \subseteq R$ for $\forall \ i = 1, 2, 3, ..., N$. Now let us define $R_{emp}(\varphi)$ as

$$R_{emp}(\varphi) = \frac{1}{N} \sum_{i=1}^{N} L_\varepsilon(\beta_i, \varphi(\alpha_i, \omega)) \tag{9.3}$$

The $R_{emp}(\varphi)$ is the empirical risk; also consider $\hat{\varphi}$ as the minimizer of the empirical risk $R_{emp}(\varphi)$ in F. The basic purpose is to use *empirical risk minimization (ERM)* theory to approximate the target function φ_0 by $\hat{\varphi}$. Now, whenever the training size N is infinitely large, the approximation of the target function can be done as $R(\varphi)$ converges to $R_{emp}(f)$ (Suykens & Vandewalle, 1999).

9.5 STRUCTURAL RISK MINIMIZATION (SRM)

Generally we face a common problem in machine learning, where we have to select a comprehensive model from the given finite data set. There are subsequent problems of over-fitting and the resulting model ending up with a strongly customized version of a previous model that creates complications. Structural risk minimization (SRM) is basically reducing the complexity of the model by stabilizing the training data. The level of difficulty of the class functions which carried out categorization or regression and the algorithm's generalizibility are related. There is always a finite set of observations in practical problems. One of the major disadvantages of ERM is that expected risk is not always minimized, which increases the experimental risk over F. Therefore, Vapnik-Chervonenkis (VC) (Vapnik & Chervonenkis, 1971) developed SRM to minimize the expected risk. The technique developed by Vapnik is similar to minimum assumption approaches. The method proposed by Vapnik and Chervonenkis (1971) (VC) gives a universal measure of difficulty and establishes limits on errors as a function of complexity. SRM is the minimization of these limits, which is based on empirical risk and capability of the class of functions.

9.6 SUPPORT VECTOR REGRESSION (SVR)

We have been using SVM as a popular tool for classification of problems in machine learning. We already know that SVMs simply separate the given data into different classes by finding the line or hyper-plane and separate the kernel supports to find the hyper-plane in the higher dimension. In regression problems, the SVR works on the same principle as SVM (i.e., using training samples, we look for a suitable function which is capable to approximate mapping from an input domain to R). The SVR trains using a symmetrical loss function with the supervised learning approach, which equally reprimands high and low misestimates. Vapnik et al. (1996) gave a generalized method to use SVM for regression problems, which we are going to discuss here in brief. Vladimir Vapnik, defines $\varepsilon-$insensitive loss function in SVR

$$L_\varepsilon(\beta, \varphi(\alpha, \omega)) = \begin{cases} 0 & if \quad |\beta - \varphi(\alpha, \omega)| \le \varepsilon \\ |\beta - \varphi(\alpha, \omega)| - \varepsilon & otherrwise \end{cases} \qquad (9.4)$$

where $\beta_i \in \mathfrak{R}$ is the output vector due to input vector α and $\varphi(\alpha, \omega)$ is a family of functions parameterized by ω. Then, to minimize the empirical risk

$$R_{emp}(\varphi) = \frac{1}{N} \sum_{i=1}^{N} L_\varepsilon(\beta_i, \varphi(\alpha_i, \omega)) \qquad (9.5)$$

The QPP is defined as minimize

$$Z(\omega, \zeta, \zeta*) = \frac{1}{2}\|\omega\|^2 + C \sum_{i=1}^{N} \left(\zeta_i + \zeta_i^*\right) \qquad (9.6)$$

subject to

$$\beta_i - \omega^T \phi(\alpha_i) - b \leq \varepsilon + \zeta_i; \ \forall \ i = 1, 2, \ldots, N \tag{9.7}$$

$$\omega^T \phi(\alpha_i) + b - \beta_i \leq \varepsilon + \zeta_i^* \tag{9.8}$$

and the non-negativity conditions are

$$\zeta_i \geq 0, \ \zeta_i^* \geq 0 \tag{9.9}$$

where vector ω is augmented with the scalar b. To solve the above equation, we shall to use Lagrange's multipliers that are defined as $\alpha = (\alpha_1, \alpha_2, \ldots, \alpha_N)^T$; also the α^* is defined as $\alpha^* = (\alpha_1^*, \alpha_2^*, \ldots, \alpha_N^*)^T$, where $\alpha_i \geq 0$, $\alpha_i^* \leq C$ and for the support vector $\alpha_i > 0$, $\alpha_i^* \leq C$. Further, the best possible hyper-plane is obtained as (Adhikari & Agrawal, 2013)

$$\beta(\alpha) = \sum_{i=1}^{N} (\alpha_i - \alpha_i^*) k(\alpha, \alpha_i) + b_{opt} \tag{9.10}$$

The time series prediction methods have been extensively used in the past few years for forecasting and giving warning for possible process malfunctioning. There are many supporting tools like SVMs, Elman recurrent neural networks and Autoregressive Moving Average (ARMA) that have been used to perform time series forecasting. In particular, to apply SVM, many algorithms have been developed in time series fore-casting like least square support vector machine (LS-SVM) algorithm (Mukherjee et al., 1995; Raicharoen & Lursinsap, 2002) and similar algorithms, Critical Support Vector Machine (CSVM) (Astuti et al., 2014) algorithm, Dynamic Least Square Support Vector Machine (DLS-SVM) (Agarwal & Kumar, 2016) and the Recurrent Least Square Support Vector Machine (Fabian & Stapor, 2017). In the next section, we discuss LS-SVM and DLS-SVM.

9.7   THE LS-SVM METHOD

To reduce computational complexity in SVM, J. A. K. Suykensand and J. Vandewalle introduced the LS-SVM classifier in 1999 (Suykens & Vandewalle, 1999). While solving quadratic problems, the inequality constraints are replaced with equality constraints in LS-SVM; this increases the training. Let us consider the input data set of N points, $\{\alpha_i, \beta_i\}_{i=1}^{N}$, where $\alpha_i \in \Re$ is input and $\beta_i \in \Re$ are the responses. Consider the optimization problem

$$\underset{\omega, b, e}{\text{Minimize}} \ J(\omega, e) = \frac{1}{2}\omega^T \omega + \frac{1}{2}\gamma \sum_{i=1}^{N} e_i^2 \tag{9.11}$$

subject to

$$\beta_i = \omega^T \varphi(\alpha_i) + b + e_i; \ \forall \ i = 1, 2, \ldots , N \tag{9.12}$$

where γ is the regularization parameter and φ is the NL mapping to a space of the higher dimension. Consider the RBF kernel $K(\alpha, \beta) = \exp\left(\frac{-\|\alpha - \beta\|^2}{2\sigma^2}\right)$, where the turning parameter σ can be employed, which satisfies Mercer's conditions (Mercer, 1909). The optimization problem for the primal space model is given by

$$\beta_i = \omega^T \varphi(\alpha_i) + b \tag{9.13}$$

Further, the optimization operation performs in dual space to avoid higher dimensionality of the weight ω as well as to reduce computational complexity (Adhikari & Agrawal, 2013).

The Lagrangian for the problem equation (9.11) is given by

$$L(\omega, b, e; \delta) = J(\omega, e) - \sum_{i=1}^{N} \delta_i \{\omega^T \varphi(\alpha_i) + b + e_i - \beta_i\} \tag{9.14}$$

where the Lagrangian's multipliers of α_i are $\alpha = [\alpha_1, \alpha_2, \ldots, \alpha_N]^T$ and $\alpha_i \geq 0; \ (\forall \ i = 1, 2, \ldots, N)$.

Not to calculate the partial derivative of L with respect to ω, b, δ_k, e_k, we shall apply the optimal conditions. To obtain a linear system of equations, we shall eliminate ω, and e_k, after equating to zero to the partial derivatives of equation (9.14).

$$\frac{\partial L}{\partial \omega} = 0 \rightarrow \omega = \sum_{i=1}^{N} \delta_i \varphi(\alpha_i) \tag{9.15}$$

$$\frac{\partial L}{\partial b} = 0 \rightarrow \sum_{i=1}^{N} \delta_i = 0 \tag{9.16}$$

$$\frac{\partial L}{\partial e_i} = 0 \rightarrow \sum_{i=1}^{N} \delta_i = \gamma e_i \tag{9.17}$$

$$\frac{\partial L}{\partial \alpha_i} = 0 \rightarrow \omega^T \varphi(\alpha_i) + b + e_i - \beta_i = 0 \quad \forall \ i = 1, 2, \ldots , N \tag{9.18}$$

Hence, the linear system obtained is

$$\begin{bmatrix} 0 & 1^T \\ 1 & \Omega + \gamma^{-1}I \end{bmatrix}_{(N+1)\times(N+1)} \begin{bmatrix} b \\ \delta \end{bmatrix}_{(N+1)\times 1} = \begin{bmatrix} 0 \\ \beta \end{bmatrix}_{(N+1)\times 1} \qquad (9.19)$$

where $\beta = [\beta_1, \beta_2, ..., \beta_N]$, $1 = [1, 1, ..., 1]$ and Ω with $\Omega(i, j) = k(\alpha_i, \alpha_j)(\forall\ i = 1, 2, ..., N)$ is the kernel matrix. Now, consider the function $\beta(\alpha)$

$$\beta(\alpha) = \sum_{i=1}^{N} \delta_i k(\alpha, \alpha_i) + b \qquad (9.20)$$

where the $\beta(\alpha)$ is the LS-SVM decision function and δ and b are the solution of the linear system of the equation (9.19).

In this way, we can convert the complex QPP system into a linear system of equations, and this will reduce computation cost as well as efforts to solve the system. We also get fast convergence rates as well as high accuracy with the LS-SVM technique.

9.8 THE DLS-SVM TECHNIQUE

Though LS-SVM is suitable for time series forecasting, to make it more convenient, it has been modified to DLS-SVM using a slightly improved computation technique.

In DLS-SVM, we can trail the nonlinear time unpredictable systems by excluding one obtainable data point whenever there is an additional observation to maintain a constant window size. By considering the previous equation

$$\chi_N = \begin{bmatrix} 0 & 1 \\ 1^T & \Omega + \gamma^{-1}I \end{bmatrix}_{(N+1)\times(N+1)} \qquad (9.21)$$

To solve equation (9.19), we obtain the inverse of equation (9.21). Now, if a new observation is associated with χ_N, then we get χ_{N+1}

$$\chi_{N+1} = \begin{bmatrix} \chi_N & k_{N+1} \\ k_{N+1}^T & k_{N+1}^* \end{bmatrix}_{(N+2)\times(N+2)} \qquad (9.22)$$

In the above matrix, the values of k_{N+1}^*, k_{N+1} are defined as

$$k_{N+1}^* = \gamma^{-1} + k(\alpha_{N+1}, \alpha_{N+1}) = \gamma^{-1} + 1 \qquad (9.23)$$

$$k_{N+1} = [1, k(\alpha_{N+1}, \alpha_i)]^T, i = 1, 2, ..., N \qquad (9.24)$$

We use matrix inversion; lemma save the computation time and calculate χ_{N+1}^{-1}

$$\chi_{N+1}^{-1} = \begin{bmatrix} \chi_N & k_{N+1} \\ k_{N+1}^T & k_{N+1}^* \end{bmatrix} = \begin{bmatrix} \chi_N^{-1} + \chi_N^{-1} k_{N+1} k_{N+1}^T \chi_N^{-1} \rho^{-1} & -\chi_N^{-1} k_{N+1} \rho^{-1} \\ -k_{N+1}^T \chi_N^{-1} \rho^{-1} & \rho^{-1} \end{bmatrix} \quad (9.25)$$

Here

$$\rho = k_{N+1}^* - k_{N+1}^T \chi_N^{-1} k_{N+1} \quad (9.26)$$

Using the above recursion equation, all direct matrix inversions are eliminated. Now, we find out χ_N^{-1}, which can be obtained while solving the equation (9.19). An additional data point has been included to remove the first point from the training data set, and χ_{N+1}^{-1} is already obtained.

Now, to obtain the new matrix system, we reorganize the set of training data $[\alpha_2, \alpha_3, ..., \alpha_{N+1}, \alpha_1]$, we obtained

$$\overline{\chi}_{N+1} = \begin{bmatrix} \overline{\chi}_N & k_1 \\ k_1^T & k_1^* \end{bmatrix} \quad (9.27)$$

Here,

$$k_1^* = \gamma^{-1} + k(\alpha_1, \alpha_1) = \gamma^{-1} + 1, \quad (9.28)$$

$$k_1 = [1, k(\alpha_1, \alpha_i)]^T, i = 1, 2, ... , N + 1 \quad (9.29)$$

$$\overline{\chi}_{N+1} = \begin{bmatrix} 0 & 1^T \\ 1 & \Omega + \gamma^{-1}I \end{bmatrix}_{(N+1)\times(N+1)} \quad (9.30)$$

$$\Omega_{ij} = k(\alpha_i, \alpha_j), i, j = 2, 3, ... , N + 1 \quad (9.31)$$

Since there is only number of rows and columns different between χ_{N+1} and $\overline{\chi}_{N+1}$, then χ_{N+1}^{-1} and $\overline{\chi}_{N+1}^{-1}$ will also have the same difference. Hence, $\overline{\chi}_{N+1}^{-1}$ can be easily obtained by adjusting the position of elements of χ_{N+1}^{-1}. Then, by using the matrix inversion formula, we get

$$\overline{\chi}_{N+1}^{-1} = \begin{bmatrix} \overline{\chi}_N & k_1 \\ k_1^T & k_1^* \end{bmatrix}^{-1} = \begin{bmatrix} \chi_{(N+1)\times(N+1)}^* & P_{(N+1)\times 1} \\ P^T & q \end{bmatrix} \quad (9.32)$$

Then, using equation (9.25) and equation (9.32), we can find

$$\overline{\chi}_N^{-1} = \chi^* - \frac{PP^T}{q} \qquad (9.33)$$

Finally, we will find δ and b from equation (9.19) with $\overline{\chi}_N^{-1}$ and repeat the iterations until we get all the data points. Hence, we can apply this technique for real-time series forecast with better performance and greater efficiency.

So, we have seen so far that SVM is a useful method, provided we have a suitable selection of kernel parameter, SVR constant, regularization constant, etc. The success of the above method is totally dependent on the proper selection of these parameters; otherwise, the resultant forecast may be absurd and unrealistic. But there is no proper method to select suitable hyper-parameters before applying the method; therefore, while applying the method, we should adopt techniques like cross-validation or Bayesian inference (Berger, 1985). We should always keep in mind that the pre-chosen values of these parameters may slightly change based on future observation.

REFERENCES

Adhikari, R., & Agrawal, R. K. (2013). *An introductory study on time series*. LAP Lambert Academic Publishing.

Agarwal, D. K., & Kumar, R. (2016, February). Spam filtering using SVM with different kernel functions. *International Journal of Computer Applications, 136*(5), 16–23.

Astuti, W., Akmeliawati, R., Sediono, W., & Salami, M. J. E. (2014). Hybrid technique using singular value decomposition (SVD) and support vector machine (SVM) approach for earthquake prediction. *IEEE Journal of Selected Topics in Applied Earth Observations and Remote Sensing, 7*(5), 1719–1728.

Berger, J. O. (1985). *Statistical decision theory and Bayesian analysis* (2nd ed.). Springer.

Campbell, M. J., & Walker, A. M. (1977). A survey of statistical work on the MacKenzie River series of annual Canadian lynx trappings for the years 1821–1934, and a new analysis. *Journal of the Royal Statistical Society: Series A, 140*, 411–431.

Cochrane, J. H. (1997). *Time series for macroeconomics and finance*. Graduate School of Business, University of Chicago.

Cortes, C., & Vapnik, V. N. (1995). Support vector networks. *Machine Learning, 20*(3), 273–297.

Cottrell, M., Girard, B., Girard Y., Mangeas, M., & Muller C. (1995). Neural modeling for time series: A statistical stepwise method for weight elimination. *IEEE Transactions on Neural Networks, 6*, 1355–1364.

Fabian, P., & Stapor, K. (2017). Developing a new SVM classifier for the extended ES protein structure prediction. *Federated Conference on Computer Science and Information Systems*, pp. 169–172.

Gareth J., & Swift, L. (1993). *Time Series forecasting simulation applications*. Ellis Horwood.

Jenkins, G. E. P., Reinsel, G. C., & Ljung, G. C. (2016). *Time Series Analysis: Forecasting and Control*, 5th edition Wiley.

Mercer, J. (1909). Functions of positive and negative type, and their connection with the theory of integral equations. *Proceedings of the Royal Society of London. Series A, Containing Papers of a Mathematical and Physical Character, 83*(559), 69–70.

Mueller, K. R., Smola, A. J., Raetsch, G., Schoelkopf, B., Kohlmorgen, J., & Vapnik, V. N. (1995). Predicting time series with support vector machine. *Machine Learning, 20*(3), 2730–297.

Mukherjee, S., Osuma E., & Girosi, F. (1995). Nonlinear predicting of chaotic time series using support vector machine. *Proceedings of the 1997 IEEE Signal Processing Society Workshop*, pp. 511–520.

Parrelli, R. (2001). *Introduction to ARCH & GARCH models*. Optional TA Handouts, Econ 472 Department of Economics. University of Illinois.

Raicharoen, T., & Lursinsap, C. (2002). Critical support vector machine without Kernel function. *Proceedings of 9th International Conference on Neural Information Processing (ICONIP)*.

Suykens, J. A. K., & Vandewalle, J. (1999, June). Least squares support vector machines classifiers. *Neural Processing Letters, 9*(3), 293–300.

Suykens, J. A. K., & Vandewalle, J. (2000, July). Recurrent least squares support vector machines. *IEEE Transactions on Circuits Systems-I, 47*(7), 1109–1114.

Vapnik, V. N. (1998). *Statistical learning theory*. Wiley New York.

Vapnik, V. N., & Chervonenkis, A. Y. (1971). On the uniform convergence of relative frequencies of events to their probabilities. *Theory of Probability & Its Applications, 16*(2), 264–270.

Vapnik, V. N., Golowich, S. E., & Smola, A. (1996). Support vector methode for function approximation regression estimation and signal processing. In *Advances in neural information processing systems 8*. CA, San Mateo.

10 A Comprehensive Review of Urban Floods and Relevant Modeling Techniques

Aadhi Naresh[1], Ravali Bharadwaj[2],
M. Gopal Naik[3], Harish Gupta[3], M. Mohan Raju[4],
and Dinesh. C. S. Bisht[5]

[1]Department of Civil Engineering, University College of Engineering, Osmania University, Hyderabad, India
[2]Department of Environmental Sciences, University College of Science, Osmania University, Hyderabad, Telangana, India
[3]Department of Civil Engineering, University College of Engineering (A), Osmania University, Hyderabad, India
[4]Nagarjunasagar Project, Irrigation & CAD (PW) Department, Government of TS, Hill Colony-508202, India
[5]Jaypee Institute of Information Technology, Noida, India

10.1 INTRODUCTION

Most natural disasters recorded in the world occur in South Asia, and they cause massive destruction to human lives and property. Due to its geographical position, climate and geological setting, India is most vulnerable to natural hazards. Among all natural hazards, India is severely affected by floods, which are the most devastating, widespread and frequent natural disasters. Excessive rainfall within a very short duration of time often triggers urban floods during monsoon season in the country. Urban flooding is a frequently occurring phenomenon in the world, and almost every city in the world is vulnerable to flooding in one way or another. The constantly growing population demands have exponentially increased the impervious surfaces in cities. Sudden heavy downpour, fewer impervious surfaces and improper sewer systems together aggravate the problem of urban floods. Generally, there is a direct interaction between surface runoff and the sewer flow system in the case of urban floods. Therefore, it is necessary to understand the phenomenon of urban flooding. Urban flooding today is a universal phenomenon, and many cities in

DOI: 10.1201/9781003102281-10

India, Europe, the United States, the United Kingdom, Australia, China and other countries have been experiencing severe floods during the last decade.

Urban flooding is one of the significant difficulties that India confronts each year, and to this point, there is no prevention available. Metropolitan cities like Delhi, Chennai, Hyderabad, Mumbai and Bengaluru and many major towns and cities across India witnessed a series of devastating urban floods in the past decade. Urban floods in the past two decades include: Chennai (2015), Srinagar (2014), Mumbai (2005), Kolkata (2013) and Hyderabad (2016) (Rangari et al., 2018). Urban areas like Mumbai, Delhi, Chennai and Hyderabad in India have battled with urban flooding for the past ten years. From the chronicled perspective, large floods in Hyderabad were seen in 2000, 2002, 2006, 2008, 2014, 2016 and 2020, which were due to significant urbanization.

Major South Asian cities like Dhaka, Islamabad, Rawalpindi, Bangkok and Colombo frequently suffer from urban flooding. In developing countries, urbanization has increased from <25% in 1970 to >50% in 2006 (UNDP-India, 2010 and NDMA). In 2001, around 285 million Indian citizens resided in 35 metropolitan cities (cities with a population above 1 million). And this statistic is projected to exceed 600 million by 2030 in 100 metropolitan cities due to the exponential trend. Global climatic variability and regional ecological challenges are notable factors which aggravate flood risks and the impact on local communities. Urban flooding was primarily a concern of municipal government and local environmental bodies but has now attained the status of "disaster", drawing the attention of disaster risk managers and environmental scientists. Unplanned urbanization is the foremost reason causing urban floods in India, which poses a dynamic challenge to city administrators, planners and dwellers. The rainfall in the urban areas is untimely, and lack of adequate drainage facilities for runoff of water creates major havoc. Diminishing urban water bodies have made the rainfall water come up into dwelling units. This phenomenon affects the densely populated urban areas with significant loss of lives and damage to public and private property.

The massive fatalities from floods across the global mega-cities have triggered actions dealing with urban floods as an important issue, which indicates that we are still trying to develop an effective response system to manage floods. This includes both structural measures and non-structural measures taken to confront the problem. Structural measures imply the construction of structures like restoring lakes, restricting encroachments and increasing the capacity of drainage channels. However, migration of a large number of people to the cities has increased day by day, and it is difficult to restrict the encroachments and reduce the city's impervious surfaces. Besides, there are various difficulties in increasing the existing drainage channels' carrying capacities and in building new storm networks. Hence, it is difficult to protect all flood-prone areas from floods of different magnitudes. As mentioned above, structural measures are not very practical due to financial constraints; hence, non-structural measures are given more importance. Non-structural measures in flood management include flood plain zoning, flood plain regulation, flood plain management and flood forecasting.

1. *Flood Plain Zoning:* Classify the areas prone to floods of different frequencies and magnitude.
2. *Flood Plain Regulation:* Regulate human activities in various flood plain zones by grouping them under different priorities. For example, public utilities like hospitals and fuel bunks are categorized under high-priority zones, and these zones should have structures (parks and green belts) built on priority to reduce the risk of disaster. Buildings like government offices and schools can be given moderate priority.
3. *Flood Plain Management:* Reduce the occupancy of the land in flood-prone areas. However, it may not be possible due to the increasing demand of habitats in the urban environment.
4. *Flood Forecasting:* Use early flood forecasts to help concerned authorities take up timely rescue measures and relief operations.

Space technology places a prominent role in flood analysis, forecasting and disaster estimation. Satellite imagery gives complete, synoptic and temporal coverage of very large areas and helps clarify the physical features and changes occurring in natural drainage systems over a time period. It gives on or near real-time spatial information on flood extent, flood monitoring and disaster assessment (Haq et al., 2012). This information is every important in water hydraulic engineering. Spatial analysis of satellite images helps to detect changes in various sections of the inundated flood plains. The multi-temporal data from the satellites are valuable in identifying the sites that are ideal to take up structural measures in controlling floods. Geographic information systems (GIS) integrate the satellite data with other secondary data, which helps prepare a detailed analysis for proposing models for flood management. GIS is an ideal tool for numerous flood-plain management activities such as base mapping, topographic mapping, mapping of flood-plain extents and depths and post-disaster verification. Remote sensing and GIS technology substitute, supplement or complement the existing flood management systems. Using these technologies extensively has a great prospect in generating a long-term database on flood-prone extent, risk assessment and relief management.

Climate data either in a statistical format or as a radar data set are essential for the prediction of floods. Accurate rainfall data, along with evaporation and runoff details, are required to make forecasts based on hydrologic models. Generally, around 30–40 years of period data are considered a time series for flood forecasting. Time series forecasting is also helpful in prediction of future possible urban floods based on previously observed values. The time series analysis has advanced to a greater extent in recent years due to upgrading of non-linear dynamics (Sparave, 1994). A model is a simplified representation of a real-world system (Sorooshian et al., 2008). The best model is the one which gives results close to reality using the fewest parameters and the least model complexity (Devia et al., 2015). Hydrologic models are primarily used for predicting the system behavior and for understanding various hydrologic processes. A model comprises various parameters which define the characteristics of the model. A runoff model is defined as a set of equations which help in estimating the runoff as a function of various parameters used in describing the watershed characteristics. The two vital inputs required in all models are rainfall data and drainage area. Other watershed characteristics like vegetation cover, soil properties, watershed topography and groundwater aquifer are also taken into consideration. Hydrologic models today are very important and essential tools in water and environment resource management. Hydraulic models are related with the dynamics of flow in the channels and over the bank areas. Hydraulic models use the boundary conditions from the hydrologic model results and the recorded flood data (Cooperative Research Centre for Catchment Hydrology, 2006).

There are different types of models—1D, 2D or 3D—based on their ability to model in 1D, 2D or 3D spaces. The 1D model takes the cross-sections to describe a channel's geometry. In 1D models, streamlines of flow are considered in intersecting the cross-sections at right angles. In reality, this is incorrect, as the velocity is present in both parallel as well as perpendicular directions of the stream. The 1D model can be represented in 2D behavior by splitting up the 1D cross-section into multiple 1D cross-sections. This is often described as quasi 2D modeling. 2D models are those that consider both x and y components of the velocity of flow. This is an approximation closer to reality as compared to the 1D model. 2D models are well suited where there are broad estuaries or wide floodplains. Generally, more data and more computational power is required to carry out a 2D model simulation. 3D models are those where all three components of velocity in x, y and z directions of flow are considered. However, for flood modeling, 3D modeling is not necessarily considered (USACE, 2002).

In the case of all stormwater modeling software, the basic formulas of EPA-SWMM are the core. A quick overview of urban floods and models are provided in Table 10.1.

10.2 DETAILED DESCRIPTION OF URBAN FLOOD MODELS

10.2.1 Storm Water Management Model (SWMM)

SWMM is a complex, coupled mathematical model and was first formulated by the US Environmental Protection Agency (EPA). SWMM can be applied not only for individual events, but also for long-period simulation of quantity of flow and efficiency in urban catchments. A simple SWMM process is shown in Figure 10.1 (Source: Computational Hydraulics International).

The SWMM runoff portion shown in Figure 10.2 (Source: Computational Hydraulics International) runs with an integrated approach which includes a group of pipe networks, open channels, management systems, pumps and control devices using non-linear reservoir methods on an interconnected system of sub-catchment areas that have rainfall and generate runoff, quantifies the pollutants; all these multi-routing operations have taken place in a single platform—SWMM—shown in Figure 10.3 (Source: Computational Hydraulics International).

By considering different, multiple time stages during the simulation process (Rossman, 2010), SWMM monitors the amount of flow and its quality produced from an individual sub-basin and the discharge, depth of flow, and quality in an individual pipe and open channel as represented in Figure 10.4 (Source: Computational Hydraulics International).

TABLE 10.1

Quick Overview of Urban Flood Models

S.No.	Name of the Model	Developer and Latest Version	Main Aim of the Model	Date of Release	Open Source?	Commercial?	Type of Model (1D,2D or 3D)	Input Data
1	SWMM (Storm Water Management Model)	USEPA Latest Version: SWMM 5.1.015, July, 2020	a. SWMM can be used for planning, design and analysis related storm water flow, integrated and sewer network, and some other urban drainage systems. b. It may also assist in simulating the hydrologic-hydraulic water quality model. c. It can be used for modeling a single event or a long-term framework of quality and runoff quantity in urban environments.	Between 1969–1971	Yes	No	1D	Regional urban conduit network and its details, channel hydraulics, elevation details of junctions, regional rainfall with continuous time series data, etc.
2	HEC-HMS (Hydrologic Engineering	US Army Corps of Engineers Latest Version: HEC-HMS 4.5, June, 2020	a. This model is relevant to:	1998 (as Hec-1)	Yes	No	1D/2D	Regional gauges temporal distribution rainfall data, Digital *(Continued)*

TABLE 10.1 (Continued)

S.No.	Name of the Model	Developer and Latest Version	Main Aim of the Model	Date of Release	Open Source?	Commercial?	Type of Model (1D,2D or 3D)	Input Data
	Centre-Hydrologic Modeling System)		• Urban flooding modeling • Flood-frequency analysis • Studies on water management • Studies relevant to the flood warning system and its planning • Reservoir system design • Flood forecasting, etc.					Elevation Model, flow data for calibration, basin losses if available, etc.
3	HSPF (Hydrological Simulation Program-FORTRAN)	USEPA	a. It includes a package for simulation of watershed hydrology and hydrology for small or large watersheds.	1980	Yes	No	1D	Topography particulars, land use/ land cover, precipitation, stream network, hydrologic data, etc.
4	DRAINS		a. This software is used in planning, designing	1998	Yes	No	1D	Pipe network, time series of different

#	Model	Developer	Year			Dimension	Description	Data requirements
		Watercom Pty Ltd, Latest Version: 2020.050, November, 2020					and quantifying the runoff in storm water network and urban catchments.By using sub-catchment with ILSAX and better using storage routing model hydrology, it can be capable of modeling drainage networks of all sizes, from small to up to 10 sq. km.	rainfall patterns, pit details and overflow route information
5	DR3M (Distributed Routing Rainfall-Runoff Model)	US Geological Survey	1972	Yes	No	1D	a. The DR3M model is a basin model that is used as an input to route storm runoff into a branched piping network and/or natural streams using rainfall. b. It is normally used to simulate small urban catchments.	High resolution of temporal precipitation, basin details, pipes, channels, reservoir and junction details, short-interval flow data, etc.
6	XP-SWMM (XP Storm Water Management Model)	XP Solutions	Since last 25 years	No	Yes	1D/2D	a. XPSWMM is used to model the flow of storm water and sewer network, as well as LID (WSUD) techniques.	Different storm events (5-yr, 10-yr, 100-yr, etc.), quantify the flow of the watershed, land use, infiltration watershed details, etc.

(*Continued*)

TABLE 10.1 (Continued)

S.No.	Name of the Model	Developer and Latest Version	Main Aim of the Model	Date of Release	Open Source?	Commercial?	Type of Model (1D,2D or 3D)	Input Data
			b. This is used to build a network-node and spatially distributed models for storm water and wastewater systems simulation.					
7	PCSWMM (Personal Computer Storm Water Management Model)	Computational HydraulicsInternational (CHI)	Supporting storm water, wastewater, watershed modeling and integrated catchment analysis, 1D-2D modeling, and much more.	1971	No	Yes	1D/2D	Digital Elevation Model, Stormwater Network, land use and land cover, soil information of the selected basin, different temporal resolution of rainfall, flow data for calibration and validation, etc.
8	MIKE URBAN+	DHI Water and Environment	a. This software is used to study urban water networks, storm water networks, collection of sewage in separate and	2019	No	Yes	2D	Digital elevation, hydraulic particulars of conduits, channels, rainfall data as provided in different time series

No.	Software	Company / Version	Description	Release			Dimension	Data Requirements
			integrated systems and urban flood modeling. b. It has characteristics that include fully hydrologic process and hydraulic water quality modeling for all areas of urban water studies.					(accumulated depth or average rainfall intensity), etc.
9	InfoSWMM	Innovyze	a. InfoSWMM is a combined hydrologic/ hydraulic, water quality simulation model with an ArcGIS-centric for urban drainage and sewer networks. b. It also provides an analysis for the planning and management of stormwater, sanitary sewer and combined sewer networks.	Version-1 released on 2004	No	Yes	1D/2D	Time series of rainfall data (hourly, daily, monthly), DEM, stormwater network and its node, elevation details, land use and land cover, meteorologic data can also feed if required, etc.
10	MUSIC	e-Water Latest Version: May 2020	a. It was founded in Australia to develop and evaluate the Water Responsive Urban Design (WSUD) framework.	Pilot Version released in March 2001	No	Yes	1D	Stormwater Network and its node details, time series of rainfall data, flow rate, meteorologic data, etc.

(*Continued*)

TABLE 10.1 (Continued)

S.No.	Name of the Model	Developer and Latest Version	Main Aim of the Model	Date of Release	Open Source?	Commercial?	Type of Model (1D,2D or 3D)	Input Data
			b. This software is used to show the efficiency in the urban environment of storm water quality control systems.					

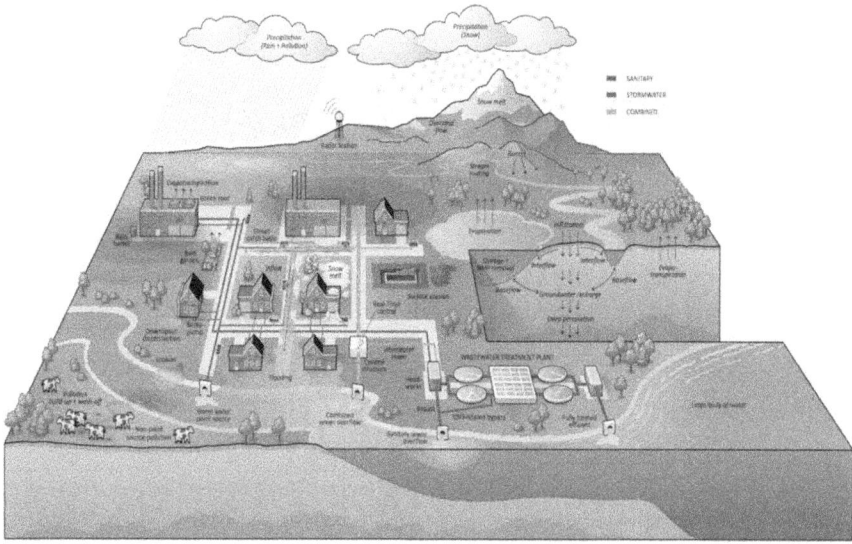

FIGURE 10.1 Basic SWMM process (*Source*: Computational Hydraulics International).

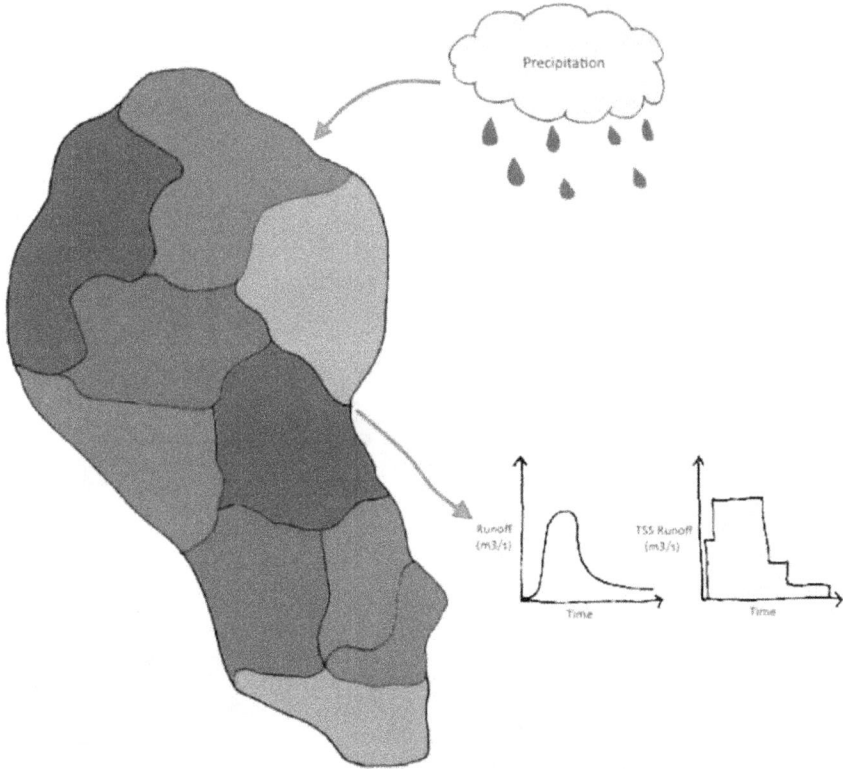

FIGURE 10.2 SWMM runoff (*Source*: Computational Hydraulics International).

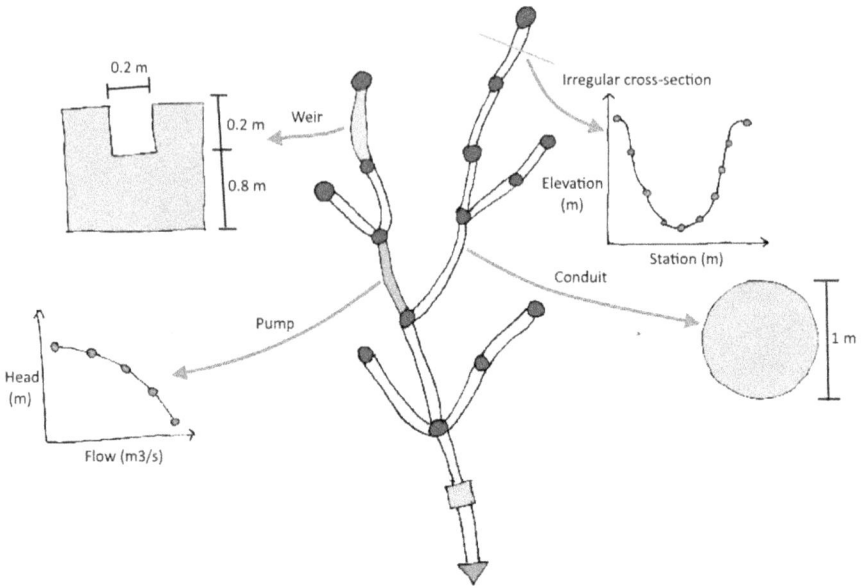

FIGURE 10.3   Flow routing process (*Source*: Computational Hydraulics International).

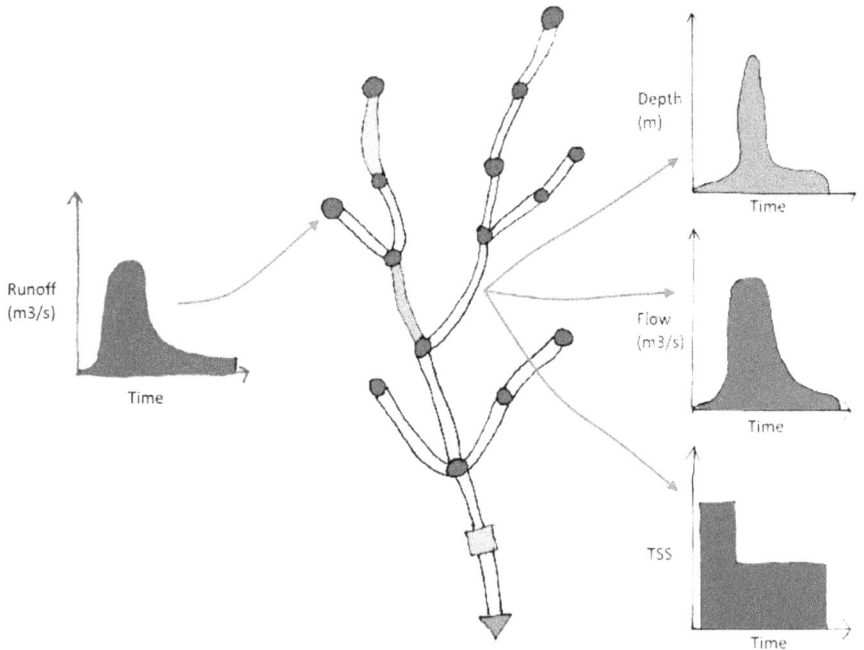

FIGURE 10.4   Hydraulic routing process with respect to discharge, depth of flow and quality in each link (*Source*: Computational Hydraulics International).

The Saint-Venant equation, the flow route for continuity and momentum calculations, is represented in the below equations:

$$\frac{\partial Q}{\partial x} + \frac{\partial A}{\partial t} = 0 \qquad (10.1)$$

$$\frac{\partial Q}{\partial x} + \frac{\partial}{\partial x}\left(\frac{Q^2}{A}\right) + gA\left(\frac{\partial y}{\partial x} - S_0\right) + gAS_f = 0 \qquad (10.2)$$

where Q is the flow rate; A represents the flow of cross-sectional area; x is the horizontal coordinate along the channel; t = duration; g = acceleration due to gravity; y = flow depth; and S_f = bottom slope of the channel.

Manning's equation in terms of flow friction of slope is represented below:

$$S_f = \frac{n^2 V^2}{R^{1.33}} \qquad (10.3)$$

where n is the roughness coefficient; V is the cross-sectional mean flow velocity and may also be written in terms of discharge and area as shown as:

$$V = QA \qquad (10.4)$$

where R = the hydraulic radius = A/P, P = wetted perimeter.

Generally, water budget consists of infiltration process, and it is a part of the equilibrium mass approach. Also, there are several methods available to measure the quantity of flow and/or the rate of penetration of water into the soil. The rate of infiltration through the unsaturated zone is calculated using Richards' equation, and another concept which is similar to Richards is finite water-content vadose zone flow strategy, which combines the absolute groundwater and surface water in homogeneous soil layers.

An easy method such as Green and Ampt's process (1911; Parlange et al., 1982) is proposed for resolving the infiltration flux for a single rainfall event in the case of uniform initial surface water content and thick, well-drained soil. There are some empirical methods such as the SCS method and Horton's method, which are well proven and used in the modeling studies. A detailed discussion about Horton's method and its usage in SWMM is presented below.

10.2.2　HORTON'S METHOD

One of the earliest and most common empirical models for simulating infiltration is Horton's equation, shown below:

$$F_p = F_c + (F_o - F_c)e^{-kt} \qquad (10.5)$$

where F_p = rate of infiltration (in./hr or mm/hr); F_c = final rate of infiltration capacity (in./hr or mm/hr); F_o = initial rate of filtration capacity (in./hr or mm/hr); t = time in hour from beginning of storm (hr); k = an exponential decay coefficient dependent on type of soil and vegetation.

10.2.3   BACKGROUND OF HORTON METHOD IN THE STORMWATER MANAGEMENT MODEL

One of the methods in SWMM available to study the infiltration process is the Horton cumulative feature protocol, and it is summarized in SWMM (Huber & Dickinsonon, 1988). Over every time interval, mean values are used. The detailed summary is mentioned below:

i. The value of f_p is a dependent factor of F, and it is calculated at each time step up to the retaining period of t_p. Finally, the mean of all infiltration potential, f_p is calculated using the below form:

$$\bar{f}_p = \frac{1}{\Delta t} \int_{t_0}^{t_1 - t_0 + \Delta t} f_p \, dt = \frac{F(t_1 - F(t_p))}{\Delta t} \tag{10.6}$$

ii. Based on the constraints, two distinct equations will then be used:

$$\begin{aligned} \bar{f} &= \bar{f}_p \text{ if } \bar{i} \geq \bar{f}_p \\ \bar{f} &= \bar{i} \text{ if } \bar{i} \leq \bar{f}_p \end{aligned} \tag{10.7}$$

where $\bar{f}$ = mean infiltration rate (length/time); i = mean rainfall rate (length/time)

iii. The cumulative infiltration at changing time periods can be written as follows:

$$F(t + \Delta t) = F(t) + \Delta F = F(t) + \bar{f} \Delta t \tag{10.8}$$

where ΔF = fΔt = gained amount of accumulative infiltration (length)
A new value of t_{p1} is found from the equation for F, and if the condition of ΔF = $f_p \Delta t$ is satisfied, then the value of t_{p1} is found simply by substituting $t_{p1} = t_p + \Delta t$. However, it is essential to work out the equation for F iterative aspect until the condition of $t_{p1} < t_p + \Delta t$ is satisfied and its process carried out with the following equations using the Newton-Raphson method.

$$FF = 0 = f_c t_p + \frac{(f_o - f_c)}{\alpha} \left(1 - e^{-kt_o}\right) - F \tag{10.9}$$

$$FF' = f_c(t_p) = f_c + (f_o - f_c)e^{-kt_o} \tag{10.10}$$

For t_{p1}, using the following equation, an initial guess is made,

$$t_{p1}(n) = t_p + \frac{\Delta t}{2} \tag{10.11}$$

In the equation (10.11), n indicates the number of iterations, and it is changed to the following form by substituting FF and FF' in the above equation:

$$t_{p1}(n + 1) = t_{p1}(n) - \frac{FF}{FF'} \tag{10.12}$$

The equation (10.13) gives a convergence condition, and it is sated below and is accomplished very rapidly.

$$\frac{FF}{FF'} < 0.001\Delta t \tag{10.13}$$

The iteration process continues until the condition of $t_p >= 16/k$ satisfies, and at $f_p = f_c$, the Horton curve becomes flat and is independent of F.

10.2.4 GREEN-AMPT METHOD

The Green & Ampt (1911) approach is physically based and can provide a clear explanation of the mechanism of infiltration. Green and Ampt (1911) suggested a fundamental physics approach that also makes assumptions leading to observational results. Infiltrated water in a saturated layer flows vertically downwards—the following simplifies the reality of penetration, as shown in Figure 10.4.

Indeed, a sharp wetting front often does not occur and/or does not saturate the soil above the wetting front. Mein and Larson (1973) modified the Green-Ampt equation, and the study considered that rate of infiltration does not change and is equal to intensity of rainfall until the time of surface ponding on the surface of the soil.

Chu (1978) separated the infiltration mechanism into two phases: phase 1, considered surface ponding, and phase 2, excluding surface ponding. Chu (1978) used the Green-Ampt equation to determine the time that separates the two processes; meanwhile, the study treats the intrusion that occurs separately at each phase as an unsteady rainfall situation.

The process of infiltration has two phases; in the first phase, the volume of water enters slowly into the soil, and at a certain stage, the soil becomes completely saturated and reaches the saturation point. At this point, we can use the Green-Ampt equation to compute infiltration capacity. However, the cumulative infiltration volume of a given soil is calculated using the equation (10.14) when the condition of cumulative infiltration volume is less than the needed cumulative infiltration volume that can induce surface saturation.

$$F_s = \frac{S.\,IMD}{\frac{i}{K_s} - 1} \tag{10.14}$$

For the condition, if $i > K_s$, $f = I$, and there is no calculation of F_s for $i < K_s$

On the other side, if the accumulated infiltration volume of a given soil is greater than the cumulative infiltration volume necessary to cause surface saturation, the following equation is used:

$$f = f_p \ and$$
$$f_p = K_s\left[1 + \frac{S.IMD}{F}\right] \tag{10.15}$$

The parameters of the above equation (10.15) with detailed description follow:

f = rate of infiltration (length/time); f_p = infiltration capacity (length/time); i = intensity of rainfall (length/time); F = cumulative infiltration capacity of the event (length); F_s = cumulative infiltration volume required to cause surface saturation (length); S = average capillary suction at the wetting front (length); IMD = initial moisture deficit for this event (length/length); K_s = saturated hydraulic conductivity of soil (length/time).

The amount of rainfall required to saturate the soil relies on the actual rainfall intensity value, and it is calculated using the equation (10.14). The surface saturates if the accumulated amount of infiltration of a given event is greater or equal to the quantity of accumulated infiltration, and for this condition, the equation 10.7, with other equations, is used to calculate the infiltration of soil. All the water infiltrates as rainfall occurs at an intensity less than or equal to the stressed hydraulic conductivity. The equation (10.15) shows that the penetration potential is dependent on the amount of water penetrated after surface saturation. The total amount of infiltration in previous periods, in particular, is dependent on the pace of infiltration. To eliminate computational errors, an integrated version of the Green-Ampt equation was developed, and its modified form is obtained as follows by substituting the infiltration force, f_p, by dF/dt and by integration.

$$K_s(t_2 - t_1) = F_2 - C - \ln(F_2 + C) - F_1 + C.\ \ln(F_1 + C) \tag{10.16}$$

where C = IMD. S (length of water); t_1 & t_2 = start time and end time, respectively.

The total infiltration of a soil is calculated with the help of the Newton-Raphson method employed in the RUNOFF block by iterately for F_2 at the end of the time interval.

There are four different cases that can occur while calculating infiltration numbers. First, if the soil's surface does not saturate over the time span (t_2-t_1), the volume of penetration is measured by multiplying the intensity of the time interval (t_2-t_1). Second, if soil saturation occurred during the previous period and ample surface water is available, the infiltration volume is estimated by the difference in accumulated infiltration amount (F_2-F_1).

In the third case, the quantities of penetration over the soil and also at this stage of the phase are estimated and summed together over the time span. The final case is the scenario where rainfall ceases or falls below the penetration potential. In this situation, it is allowed to infiltrate any water placed on the surface and add to the overall amount of absorption.

10.2.5 HEC-HMS (HYDROLOGIC ENGINEERING CENTRE-HYDROLOGIC MODELING SYSTEM)

This tool was selected for the United States and designed to simulate the hydrologic reaction of dendritic watershed systems (USACE, 2000) by the Army Corps of Engineers, which is commonly used by hydrologists and engineers in numerous scientific studies. The hydrologic and hydraulic modeling method components are divided into three categories: components of the basin, meteorologic and time series. In addition, to link the sub-basins, certain parts of hydraulic components such as conduits or reservoirs are made available. In this model, multiple options are available for model penetration (Horton, Green-Ampt) where only minimal options are available for conduit flow modeling (the kinematic wave equation). Additional elements have been integrated into HEC-HMS, such as complexity analysis, sediment movement, erosion modeling and nutrient-related water quality (USACE, 2016). Older HEC-HMS models consisting of input/output data might be somewhat bulky. The new update has brought major changes to the GUI (Halwatura & Najim, 2013). Additional knowledge on this model is provided in Table 10.1.

Generally, the model contains conceptual and various loss estimation methods, and they are the transformation of rainfall-runoff and routing system. SCS-CN, Snyder unit hydrograph, SCS-unit hydrograph and Clark hydrograph are the drainage models concerned.

10.2.6 SCS CURVE NUMBER METHOD

SCS curve number method is an empirical method utilized in different hydrologic studies for calculation of runoff or infiltration with rainfall as input. This method is widely used in all hydrologic studies for estimation of direct runoff from the occurrence of rainfall in a particular area. The curve number depends on the regional hydrologic soil community, land use, treatment and hydrologic condition. Research from many small experimental watersheds led to the creation of an empirical formula, shown below:

$$I_a = 0.2S \quad (\lambda = 0.2, \quad \text{for India}) \tag{10.17}$$

On this basis

$$P_e = \frac{(P - 0.2S)^2}{(P + 0.8S)} \tag{10.18}$$

where I_a = initial abstraction; P_e = rainfall excess; P = total precipitation

$$S = 254\left(\left(\frac{100}{CN}\right) - 1\right)$$ (10.19)

With the values of P and P_e from many watersheds, plots were developed and used in defining the SCS normal curves. So, for the standardization of these curves, a non-dimensional curve number CN is defined (0 < CN < 100). Generally, the value of CN is not uniform and varies depending upon the nature of the surface condition. So, for an impermeable condition, the CN value was assigned as 100 and for natural surfaces it would be CN < 100.

10.2.7   LIMITATIONS OF THE APPLICATION

No expression of time is found in the equation. It is designed to predict runoff from single storms. There is some regeneration of penetration rates during the cycles of a discontinuous storm that has intervals of no rain. A fresh, higher CN is typically chosen on the basis of the difference in previous moisture when the rainless intervals are over an hour. The initial abstraction principle "Ia" is composed of different variables such as interception, initial absorption, surface holding capacity as well as other parameters. On the basis of evidence from both large and small watersheds, the relationship between Ia and S was calculated. Ia's further improvement is feasible, but it has not been suggested because very little is understood about the magnitudes of interception, penetration and surface storage under normal field conditions.

10.2.8   SNYDER UNIT HYDROGRAPH

The Snyder method uses variables to define the below parameters, which would then be used in the unit graph preparation since the most important parameters that characterize a unit hydrograph are peak flow and time to peak (Snyder, 1938).

$$t_p = 5.5t_r$$ (10.20)

$$p = CC_t(LL_c)^{0.3}$$ (10.21)

where C_t = basin coefficient; L = mainstream length from the outlet to the dividing point; L_c = the length of the mainstream from source to the point close to the centroid of watershed; and C = conversion constant (0.75 in SI).

   i. In this strategy, the basic principle is that basins with identical physiographic characteristics found in the same region would have similar C_t and C_p values.
  ii. It is also recommended that the basin be close or identical to the gauged basins for which the required coefficients can be determined in the case of ungauged basins.

Based on the calibration, the parameters of C_t and C_p are best defined, but these are not physically dependent parameters. Bedient and Huber (1992) state that C_t usually ranges from 1.8 to 2.2, but it was found to range from 0.4 in the Gulf of Mexico's 6 mountainous regions to 8.0. They stated that C_p ranges between 0.4 and 0.8, but higher values of C_p are associated with lower values of C_t.

10.2.9 SCS Unit Hydrograph Model

For a large range of small agricultural catchments across the United States arising from measured rainfall and drainage, the model is based on UH averages. The SCS unit hydrograph is a non-dimensional model, but a relation is developed between unit hydrograph of only one peak and time to the peak, as shown below:

$$U_p = C \frac{A}{T_p} \qquad (10.22)$$

where A = mean area of the watershed; and C = a conversion constant value (2.08 in SI).

The time to peak with a length of unit excess precipitation is compared, and it is shown below as:

$$T_p = \frac{\Delta t}{2} + t_{lag} \qquad (10.23)$$

where Δt = the excess precipitation time period (computational time interval in the run); and t_{lag} = basin delay; it is defined as the variation of time between the center of mass of excess rainfall and unit hydrograph peak.

10.2.10 Estimation of Model Parameters

SCS-UH lag can be measured by its calibration for metered headwater sub-watersheds. For ungauged watersheds, a relation is suggested by SCS-UH between UH delay duration and concentration-time, t_c, as represented below:

$$t_{lag} = 0.6 t_c \qquad (10.24)$$

Time of concentration is a dependent function, particularly on the concept of physical and mathematical principles, and could be measured as:

$$t_c = t_{sheet} + t_{shallow} + t_{channel} \qquad (10.25)$$

where t_{sheet} = Σ Travel period over the watershed surface of the layer of flow segments; $t_{shallow}$ = Σ Travel period in segments of shallow flow, down streets, gutters, or in shallow stream and gullies; and $t_{channel}$ = Σ Travel time in channel structures.

With the cross-section details of open channels obtained from field surveys, maps or aerial photographs, velocity is estimated by Manning's equation:

$$V = \frac{CR^{2/3}S^{1/2}}{n} \tag{10.26}$$

where V = mean velocity; R = hydraulic radius (it is the ratio of the area of channel cross-section to the wetted perimeter); S = slope of the energy gradient line (often approximated as channel bed slope); C = conversion constant (1.00 for SI and 1.49 for foot-pound system.); and n = Manning's roughness coefficient.

Channel travel time is computed using the above equation (10.26) once velocity is estimated:

$$t_{channel} = \frac{L}{V} \tag{10.27}$$

Generally, sheet discharge types flow across the top of the watershed land until a channel is entered by water and travels the distance in the range of 10–100 meters. The travel period of sheet-flow is represented as:

$$t_{sheet} = \frac{0.007\,(NL)^{0.8}}{(P_2)^{0.5}S^{0.4}} \tag{10.28}$$

where N = coefficient of overland flow roughness; L = flow length; P_2 = 2-year, 24-hour rainfall depth, in inches; and S = hydraulic grade line, which may be approximated by the slope of the land.

For different surfaces, N values are different, and these are given in Table 10.2. Sheet water flow normally converts to concentrated shallow water flow after 100 meters.

$$V = \begin{cases} 16.1345\sqrt{S} \ for \ unpaved \ surface \\ 20.3282\sqrt{S} \ for \ paved \ surface \end{cases} \tag{10.29}$$

10.2.11   HSPF (HYDROLOGICAL SIMULATION PROGRAM-FORTRAN)

The HSPF model is a robust package that can simulate the constant, dynamic or steady-state conditions of all hydrologic/hydraulic and water-quality processes in a watershed. It was developed by the Environmental Protection Agency (EPA). Its impermeable land module combines hydrologic and water-quality methodology into natural and human-induced water sources and urban watersheds.

It is the only systematic model that has applications in water supply systems planning, construction and service. This model also provides the possibility of using probabilistic analysis by its continuous simulation capabilities in the area of

TABLE 10.2

Overland-flow Roughness Coefficients for Sheet-flow Modeling (Values Given by USACE, 1998)

Surface Description	N
Smooth surfaces (concrete, asphalt, gravel, or bare soil)	0.011
Fallow (no residue)	0.05
Cultivated soils:	
Residue cover <20%	0.06
Residue cover >20%	0.17
Grass:	
Short grass prairie	0.15
Dense grasses, including species such as weeping love grass, bluegrass, buffalo grass, blue grass, and native grass mixtures	0.24
Bermudagrass	0.41
Range	0.13
Woods[1]:	
Light underbrush	0.40
Dense underbrush	0.80

Note: [1]In this case, consider plant cover about 0.1 ft, and it is the only scenario where plant cover obstructs the sheet water flow.

hydrology and water quality control. The input data used to model the watershed phase are: historic rainfall data, temperature, amount of evaporation and parameters relevant to different aspects of land use, soil properties and agricultural type.

The other kind of data required for the model are climate data which include the amount of precipitation, evapotranspiration, air temperature, wind velocity, solar radiation, dew-point temperature and cloud cover information that is feeded into the model on an hourly basis.

Based on the input data fitted into the model, different outputs are extracted. One output results in the quantity of water present in the ground water aquifer. In this case, the input is time history of the volume of water transferred across the soil's surface. This model can also estimate the flow rate of runoff, sediment rate, nutrients, contaminants, hazardous chemicals and other water quality levels (Duda et al., 2012).

10.2.12   DRAINS

DRAINS is a stormwater drainage system design and analysis Windows program widely used in Australia. This software can be used to model small to very large areas. The model is set up using multiple sub-catchments with April 2016 and

ILSAX hydrology for an area up to 10 km², and for extensive areas, the storage routing model is used.

During the period of a flood condition, it has the potential to operate on various time measures, translates rainfall variations to hydrographs for stormwater drainage and routes them across networks of drains, storage basins, rivers and streams. In this framework it incorporates:

    i. Design and analysis operations
    ii. Five alternative hydrology models and also two different hydraulics methods
    iii. Closed piped flow and open channel flow systems
    iv. Retaining wall, culverts and other structures
    v. Stormwater detention structures
    vi. Urban and rural catchments of a very large scale

With its single package of DRAINS, ARR2016, ILSAX can perform stormwater drainage system design and analysis, hydrologic modeling, rational approach, storage routing modeling and also unsteady hydraulic modeling of piping networks, open channels and surface overflow routes in the scenario of the premium hydraulic model. This software also includes an automated design process for piped drainage systems, CAD and GIS software connections, and an integrated support system.

10.2.13 APPLICATIONS

This software covers all fields of the watershed system, mentioned below.

    i. Urban flooding
    ii. Drainage asset modeling
    iii. Subdivision drainage streets
    iv. Infill design property drainage and onsite stormwater detention (OSD)
    v. Detention basins
    vi. Highway drainage
    vii. Rural-urban catchment modeling
    viii. Channel hydraulics

Each model has some disadvantages, such as DRAINS still having some drawbacks, but it does not have (a) consistent wet and dry conditions modeling for long periods, (b) modeling of water quality, and (c) 2D unsteady state of flow modeling.

10.2.14   DR3M (DISTRIBUTED ROUTING RAINFALL-RUNOFF MODEL)

This tool is often stormwater modeling software and is designed to simulate the runoff of storms. This software can be used to develop a model of conduit or natural channel routing by using rainfall as an input as per the time frame chosen by the user. DR3M is typically used to simulate flood runoff for small urban areas and uses the Green-Ampt equation to quantify infiltration and rainfall excess in the previous area.

This software's biggest drawback is that it can not model the basin's interflow and baseflow. A short duration discharge is required. For the model's optimization and calibration, a short duration discharge is needed (Survey USG, Summary of DR3M). The main input data required for this model are: i) daily rainfall, ii) daily evapotranspiration scenario and iii) short-duration precipitation.

10.2.15 XP-SWMM (XP Storm Water Management Model)

This software is a dynamic simulation program, and it helps in doing hydraulic, hydrologic modeling and also sustainable drainage systems planning and management. Besides, it has a GIS interface and is used in standard LID (WSUD) schemes to mimic stormwater and drainage flows as well as treatment. Using the 1D (channel and pipes) –2D (surface grid) concept, flows are simulated, and maps are generated; it will provide information on the extent of flooding and depth.

Meanwhile, this model can also help perform the analysis of natural rainfall-runoff processes and also control the water supplies using certain engineered structures. Another phase of XPSWMM is that it can integrate stormwater and wastewater systems design, analysis and simulation. XPSWMM has also been used to model and simulate flow in natural groundwater involvement environments, including rivers, reservoirs and floodplains (XP Software, 2013).

10.2.16 Personal Computer Storm Water Management Model (PCSWMM)

PCSWMM is also another form of SWMM commercial software, and it is a third-party platform for SWMM. This software of its inbuilt program was developed by Computational Hydraulics International (CHI) in cooperation with the EPA. PCSWMM is a kind of graphical decision support system based on the SWMM program and deals with a wide variety of things, such as control of files, time series, model creation, calibration and validation. Besides, an interface of GIS and its attributes has been added to the latest PCSWMM update. This software includes six major components: precipitation, runoff, temperature, storm-water network, sewerage network and storage. It also has one additional tool such as watershed delineation, which helps in dividing the sub-catchments into urban areas based on infiltration and land use. In this software, widely used standard infiltration methods such as Green Ampt, SCS curve number and Horton equation are included to estimate the amount of sub-catchment infiltration. Basically, in PCSWMM, an integrated approach of three modes of the flow routing method was incorporated: steady flow, Kinematic flow and dynamic flow routing.

Using a standard Manning's formula with additional parameters, a relation is developed between cross-sectional area, discharge, gradients and hydraulic radius, as shown in equation (10.30).

$$Q = B\frac{C_m}{n}S^{1/2}(y - y_d)^{5/2} \quad \text{(Dynamic Equation)} \tag{10.30}$$

Considering the input of rainfall rate in the controlled space volume, including the plane surface, the continuity and dynamic relation is made available between runoff (Q) and infiltration (f), given the unit width of the catchment. The equation is provided below:

$$L = \left(fL + \frac{Q}{B}\right) + L\frac{\Delta_y}{\Delta_t} \text{ (Continuity Equation)} \qquad (10.31)$$

where L = overland flow length; B = catchment breadth; C_m = 1.0 for metric units = 1.49 for imperial or US customary units; n = Manning roughness coefficient; y = water department; y_d = surface depression storage depth and f = infiltration.

10.2.17   APPLICATIONS

   i. Applies to both urban/rural watersheds, for master planning/site design purposes.
  ii. Floodplain mapping, flood hazard/risk assessment, floodplain management.
 iii. Improved dual drainage system modeling of interconnected surface overflow channels/storage areas (major system) + pipe network (minor system).
  iv. Allows distributed hydrology (i.e., gridded rainfall on sub-catchments with a very fine spatial level of detail).

PCSWMM can support and carry out drainage studies, water supply and green infrastructure planning, flood plain delineation, sewer runoff reduction, water quality and comprehensive catchment evaluation, 1D-2D modeling and many others.

PCSWMM software needs following data and it will help in improving the storm water model analysis:

  ○ Digital Elevation Model (DEM)
  ○ Land Use and Land Cover
  ○ Soil information
  ○ Storm water network with detailed node information (will help for detailed node to node analysis and overflow of manholes)
  ○ Rainfall, etc.

10.2.18   MIKE URBAN+

DHI MIKE URBAN+ is commercial software that has been developed to model urban water services, complex water management systems, stormwater runoff systems, separate and integrated sewage collection and urban flooding systems. It also facilitates hydrologic and hydraulic simulation, including the quality of water relevant to urban studies.

This software includes an advanced stormwater drainage flood modeling framework based on a modular mesh, GPU and multiple cores engines and water exchange between the stormwater network and 2D surface. It has an advanced high-performing MIKE 1D multi-core engine for modeling stormwater and sewage systems with fast execution, extracting model data directly from a high-performing database. With the latest MIKE URBAN+ high-performance interface, a user can perform daily routines speedily with hydraulic engines utilizing multiple cores and GPU—which promises accuracy and speed even when modeling mega-cities.

10.2.19   InfoSWMM

InfoSWMM is a product of Innovyze and is full commercial software. This tool is an ArcGIS-incorporated, technologically sophisticated and robust model of hydrologic, hydraulic and water quality for urban stormwater runoff and sewage systems management. It is an entirely interactive, geospatial software framework that can build a model, and it involves the hydrologic cycle process.

It has built-in ArcGIS component technology and easily combines sophisticated simulation and optimization features for hydrologic and wastewater collection systems with allowing users to optimize current GIS investments in a full and comprehensive package.

10.2.20   MUSIC (Model for Urban Stormwater Improvement Conceptualization) Software

MUSIC is a quick, leading platform and personalized workflow from Australia, making it easy to develop and evaluate Water Sensitive Urban Design (WSUD) structures and follow best practices. This tool is used in Australia by environmental architects, planners, local authority engineers and planning approval authorities to manage the effect of urban development and other improvements in land use on water resources. In Australia, MUSIC has been mandated by several city councils and states to design large-scale economic construction. To find the safest way to catch and manage stormwater runoff, eliminate its pollutants and reduce runoff frequency, MUSIC is the best tool available to model it. These processes will be tested and help meet WSUD's integrated water cycle management (IWCM) targets with the use of the MUSIC platform.

In the case of all stormwater modeling software, the basic formulas of the EPA-SWMM engine are the core heart and run through these.

10.3   DISCUSSION AND GAPS IN URBAN FLOOD STUDIES

Flooding in urban areas has become a more common phenomenon in recent years. Most Indian cities are prone to these floods and require a more scientific approach to identify these flood areas for proper management. Conventional modeling approaches (1D and 1D-1D) quite accurately simulate the drainage network. However, in the case of significant rainfall events, these models cannot simulate the inundation depth in a built-up area and visualize flood extent (Bisht et al., 2016). For estimating depth and extent in urban floods, more accurate models other than 1D and 1D-1D are needed. Basically, 1D-2D models are based on the 2D-shallow water equations and are solid tools for simulation of the floods in urban areas (Leandro et al. 2009). Most of the 2D modeling approaches available today help identify flood-vulnerable areas for critical rainfall events. Limitations in urban flood modeling include the unavailability of the extensive raw data sets, which has made urban flood modeling a more complicated process.

A study carried out by Rangari et al. (2019) on inundation risks on urban flooding for one part of the Hyderabad city used HEC-RAS for flood analysis. HEC-RAS is a freely available 2D hydraulic model that integrates with GIS and generates a depth of the flood inundation over the underlying terrain and generates risk maps for different rainfall scenarios to generate the inundation depths. Past rainfall events were used to identify the peaks and run simulations for the 5-, 10- and 15-year return periods. The model was not validated due to the lack of flow data, and also no gauges are installed in urban storm flows to monitor the flow. The study concluded that HEC-RAS was more reliable in an urban environment with limited data.

Another study carried out by Tushar Surwase and Manjusree (2019) compared flood modeling of the same area using two models, HEC-RAS and PCSWMM. They concluded that SWMM is suitable for cities where a systematic sewage network is available along with closed conduits and a stormwater drainage network.

Work carried out by Kourtis et al. (2017) has used 1D-1D and 1D-2D urban flood models such as MIKE URBAN and SWMM. This study presented a comparison of a 1D-1D urban flood model (SWMM) with 1D-2D (MIKE URBAN-MIKE FLOOD) to demonstrate the model structure and its significance. The two models were used for flood simulation of a small urban catchment (Zone D) in Athens. The simulation was done for return periods of ten and 25 years and one-hour rainfall duration. These results revealed that the 1D-1D model is faster than the 1D-2D model, but it could not simulate the flood extent and flood inundation accurately. 1D-2D modeling requires more research and also needs the implementation of rainfall-runoff monitoring for calibration and validation of the model. Secondly, sensitivity analysis for the 1D-2D model is in progress.

Akhter and Hewa (2016) used PCSWMM to explore the hydrologic responses in the Myponga catchment, as a result of land-use land cover changes and possible adaptation of water sensitive urban design (WSUD) technologies in managing floods. The calibration and validation of the model satisfactorily predicted the measured data with accuracy and reliability. Zope et al. (2015) studied the urbanization effects on flooding for Mumbai city. They used HEC-HMS based hydrodynamic modeling for this purpose and also used the soil conservation service-curve number (SCS-CN) method to estimate the loss; the SCS-unit hydrograph method was used for transformation, and the kinematic wave method was used for flood routing. Daily rainfall over a 100-year return period was taken as input for the model. Peak discharges at each of the junction nodes and at the outlet node were calculated for LULC patterns that existed during 1966 and 2009.

10.4 SUMMARY AND CONCLUSIONS

This chapter attempts to provide a brief idea of the available current urban flood modeling tools with respect to their strengths and limitations for modeling floods in urban areas. There are numerous tools available for urban flood modeling, each with its own benefits and limitations for applications in different urban set-ups. The approach followed in rapid flood spread models is simple to use, allows for minimal data and has a fast run time, but only the final inundation is the outcome. 1D sewer outputs incorporate the positions of overflow and the full volume of floodwater. Furthermore, it helps to measure the overall inundation level and floodwater depth by simplifying the overflow situation using a simulated reservoir. However, as it is incapable of identifying the surface flow, real flood conditions cannot be defined.

The dynamics of the floodwater cannot be given by flood simulation techniques for surface flow. Nonetheless, 1D surface effects are restricted to the profile of the surface network and 1D flow velocity, and further calculation time and detailed data are needed for the 2D approach. In addition, the effect of the flow in the stormwater runoff system is not considered for all methods, and the quality of the output information may be impaired. In particular, the method of sewer-surface coupling allows the urban dual-drainage system to be portrayed, as it recognizes the flow interchange between major and minor systems and gives detailed explanations of flood conditions. However, when the flow over-tops the specified surface networks, the 1D-1D method does not provide details about surface flow velocities. On the other hand, while the 1D-2D coupling methodology will provide the most reliable and thorough results, it is computationally costly in terms of both run time and data specifications. Finally, this chapter gives a comprehensive understanding of current modeling approaches' key features, such as abilities in representing the prevailing flood systems, input data requirements, future prediction information and model run time. This thorough understanding of the flood models in flood risk management will help in urban flood management and also help to carry out successful modeling tasks suitable for their criteria.

REFERENCES

Akhter, M. S., & Hewa, G. A. (2016). The Use of PCSWMM for Assessing the Impacts of Land Use Changes on Hydrological Responses and Performance of WSUD in Managing the Impacts at Myponga Catchment, South Australia. *Water*, *8*(11), 511. https://doi.org/10.3390/w8110511.

Bedient, P. B., & Huber, W. C. (1992). *Hydrology and Flood Plain Analysis,* 2nd edition. Addison-Wesley, New York.

Bisht, D. S., Chatterjee, C., Kalakoti, S., Upadhyay, P., Sahoo, M., & Panda, A. (2016). Modeling urban floods and drainage using SWMM and MIKE URBAN: A case study. *Natural Hazards*, *84*(2), 749–776.

Chu, S. T. (1978). Infiltration during an unsteady rain. Water Resources Research, *14*(3), 461–466.

Cooperative Research Centre for Catchment Hydrology. (2006). *Series on model choice: General approaches to modelling and practical issues of model choice.* http://www.toolkit.net.au/pdfs/MC-1.pdf

Devia, G. K., Ganasri, B. P., & Dwarakish, G. S. (2015). A review on hydrological models. *Aquatic Procedia*, *4*, 1001–1007. https://doi.org/10.1016/j.aqpro.2015.02.126.

Duda, P. B., Hummel, P. R., Donigian, A. S., & Imhoff, J. C. (2012). BASINS/HSPF:Model use, calibration, and validation. *American Society of Agricultural and Biological Engineers*, *55*(4), 1523–1547.

Green, W. H., & Ampt, G. (1911). Studies of soil physics, part I – the flow of air and water through soils. *Journal of Agricultural Science*, 4, 11–24.

Halwatura, D., & Najim, M. M. M. (2013). Application of the HEC-HMS model for runoff simulation in a tropical catchment. *Environmental Modelling & Software*, 46, 155–162. https://doi.org/10.1016/j.envsoft.2013.03.006

Haq, M., Akhtar, M., Muhammad, S., Paras, S., & Rahmatullah, J. (2012). Techniques of Remote Sensing and GIS for flood monitoring and damage assessment: A case study of Sindh province, Pakistan. *Egyptian Journal of Remote Sensing and Space Sciences*, *15*(2), 135–141. https://doi.org/10.1016/j.ejrs.2012.07.002.

HEC-HMS Technical Reference Manual. https://www.hec.usace.army.mil/confluence/hmsdocs/hmstrm/surface-runoff/scs-unit-hydrograph-model.

Huber, W. C., & Dickinson, R. E. (1988). Storm water management model; version 4 User Manual. https://www.chiwater.com/Company/Staff/WJamesWebpage/original/homepage/Research/T52Kipkie/T52.html.

InfoSWMM User Manual.

Kourtis, I. M., Bellos, V., Tsihrintzis, V. A., & Flood, M. U. (2017). Comparison of 1D-1D and 1D-2D urban flood models, no. September, *Proceedings of the 15th International Conference on Environmental Science and Technology (CEST 2017)*, Rhodes, Greece, 31 August to 2 September 2017, 31.

Leandro, J., Chen, A. S., Djordjević, S., & Savić, D. A. (2009). Comparison of 1D/1D and 1D/2D Coupled (Sewer/Surface) Hydraulic Models for Urban Flood Simulation. *Journal of Hydraulic Engineering*, *135*(6), 495–504.

Mein, R. G., & Larson, C. L. (1973). Modeling infiltration during steady rain. *Water Resources Research*, *9*(2), 384–394.

Parlange, J. Y., Lisle, I., Braddock, R. D., & Smith, R. E. (1982). The three-parameter infiltration equation. *Soil Science*, *133*(6), 337–341.

Rangari, A. V., Umamahesh, N. V., & Bhatt, C. M. (2019). Assessment of inundation risk in urban floods using HEC RAS 2D. *Modeling Earth Systems and Environment*, *5*(4), 1839–1851. https://doi.org/10.1007/s40808-019-00641-8.

Rangari, V. A., Umamahesh, N. V., & Patel, A. K. (2018). Development of different modeling strategies for urban flooding: A case study of Hyderabad city. *Proceeding of National Conference: Civil Engineering Conference- Innovation for Sustainability (CEC – 2018)*. 09–10th September.

Rossman L. A. (2010). Storm water management model user's manual. USEPA Software.

Snyder, F. F. (1938). Synthetic unit-graphs. *Transactions, American Geophysical Union,* 19, 447–457.

Sorooshian, S., Hsu, K., Coppola, E., Tomassetti, B., Verdecchia, M., & Visconti, G. (2008). Hydrological Modelling and the Water Cycle: Coupling the Atmospheric and Hydrological Models, ISBN: 978-3-540-77842-4.

Sparave, J.(1994). Linear neighbourhood evolution strategy. In Proceedings of the 3rd Annual Conference on Evolutionary Programming, World Scientific, Singapore, pp. 42–51.

Surwase, T., & Manjusree, P. (2019). Urban flood simulation – A case study of Hyderabad City. National Conference on Flood Early Warn. Disaster Risk Reduction, Hyderabad, India, 30–31 May 2019, no. June, pp. 133–143.

Survey USG, Summary of DR3M. USGS science for a changing world. http:// water.usgs.gov/cgi-bin/man_wrdapp?dr3m.

UNDP-India (2010). *Panel discussion on urban floods in India (Background note).* UNDP-India and NDMA, Government of India, E-circulation.

USACE (1998). *HEC-HMS Technical Reference Manual.*Hydrologic Engineering Center Davis.

USACE (2000). *HEC-HMS hydrologic modeling system user's manual.* Hydrologic Engineering Center Davis.

USACE, (2002). *HEC-HMS hydrologic modeling system user's manual.* Hydrologic Engineering Center Davis.

USACE (2016). *HEC-HMS hydrologic modeling system user's manual.* Hydrologic Engineering Center Davis.

Wheater, H., Sorooshian, S., & Sharma, K. D. (2008). *Hydrological modelling in arid and semi-arid areas* (p. 223). Cambridge University Press. ISBN-13 978-0-511-37710-5.

XP Software (2013). Technical descriprion XP solution.

Zope, P. E., Eldho, T. I., & Jothiprakash, V. (2015). *Impacts of urbanization on flooding of a coastal urban catchment: A case study of Mumbai City, India,* 887–908. https:// doi.org/10.1007/s11069-014-1356-4.

11 Fuzzy Time Series Techniques for Forecasting

Alisha Mittal and Baishali Mishra
CCS HAU, Hisar

11.1 INTRODUCTION

Various forecasting procedures such as neural networks, fuzzy logic and ARIMA are available to deal with problems related to environmental and financial time series. Forecasting is not an easy process as it includes economic, political, financial, social and many other phenomena. The variables which are characterized by approximation, imprecision, nonlinearity, vagueness, etc., are hard to measure. Moreover, the data may have challenging patterns and behaviors which lead to difficulty in performing analysis and prediction. These circumstances have given rise to new forecasting methods providing a high degree of prediction and running at low cost. One such method is fuzzy time series (FTS), which works on the concept of fuzzy logic. In recent years, this method has been highly noticed and has acquired more significance as many studies have reported its high-quality precision over the other models (Singh, 2017). However, due to methodological problems, the FTS method has suffered disapproval in the literature, for example, see Sadaei et al. (2016). Nonetheless, all forecasting problems are conditioned upon various types of doubts, concerning ambiguity in the data because of measurement errors, or other unknown factors, or concerning the fact that all features of the underlying process might not be modeled correctly, which leads to error in forecasted results. Due to these facts, two methods—probabilistic forecasting and interval forecasting—have been developed by Gneiting and Katzfuss (2014) and Chatfield (1993), respectively. Instead of estimating unique point forecasts, these two methods are concerned with estimating the distributions of certain values. The FTS models expand time series along with the use of the concept of fuzzy sets by recognizing relationships and patterns between these fuzzy sets. However, the majority of papers on FTS generate a point forecast by defuzzifying the forecasted values. Concerning the literature of FTS, the procedures to generate interval forecasts are based on type 2 fuzzy sets; see Bajestani and Zare (2011) and Huarng and Yu (2005). The major limitation of these techniques is that the type 2 fuzzy sets require high computational cost.

Therefore, this chapter focuses on the concept of fuzzy logic and FTS for forecasting problems.

11.2 FUZZY SET THEORY

Zadeh (1965) created fuzzy set theory. Since its creation, it has achieved success in both theory and application. The impulse behind fuzzy set theory is primarily to supply a proper, stronger and quantitative structure to deal with ambiguity in individual understanding by expressing it in normal language (Dubois & Prade, 1991). We encounter many situations in real life where inclusion and exclusion from a set are not clearly defined; for instance, the classes of tall boundaries of such sets are indistinguishable, and the transition from one member non-member appears slow rather than quick. A fuzzy set is defined as any set that permits its members to have dissimilar grades of membership (membership function) in the interval [0, 1]. The mathematical description is shown below:

Let X be the collection of objects, then A is a fuzzy set of ordered pairs contained in X such that

$$A = \{(x, \ \mu_A(x)): \ x \in X, \ \mu_A(x): \ X \to [0, \ 1]\}$$

where $\mu_A(.)$ is the membership function of A which is defined as a function from X into [0, 1]. The membership function provides a grade of membership of the element to the set. Fuzzy set has a graphical depiction which states how the transition from one to another takes place. Such type of graphical depiction is also known as a membership function.

11.3 FUZZY SYSTEM (FS)

A fuzzy system (FS) is defined as any fuzzy-based system which utilizes fuzzy logic as the source for information representation by using different forms of knowledge. We can model variables, interactions, systems and intermodal relationships by using many ways like membership function, shape analysis of membership function, etc. Membership functions are often used as the mathematical means of representing values for inference mechanism.

11.4 MEMBERSHIP FUNCTION

A membership function is defined as the function which assigns values to the elements contained in a universal set that fall inside a specific range and point to the membership grade of these elements in the set. The larger the values, the larger the degrees of set membership. Such type of set defined by membership functions is known as a fuzzy set. The most generally used range of values for membership function is the unit interval i.e. [0, 1].

Note: For fuzzy set *A*, its membership function is represented by μ_A

$$\mu_A: X \to [0, 1]$$

Writing it in another way, if the function is represented by A, then it has the same form

$$A: X \to [0, 1]$$

11.5 SOME FEATURES OF A MEMBERSHIP FUNCTION

The features of a membership function are described below. The graphical representation is shown in Figure 11.1.

Core: It is defined as the region identified by full membership in set A, i.e., $\mu(x) = 1$.

Boundary: It is defined as the region identified by partial membership in set A, i.e., $0 < \mu(x) < 1$.

Support: It is defined as the region identified by non-zero membership in set A, i.e., $\mu(x) > 0$.

11.6 FUZZY LOGIC

Fuzzy logic is defined as a multi-valued logic that permits intermediary values to be labeled between traditional evaluations like high/low, yes/no, true/false, etc. FS translates these rules into their arithmetic equivalents. In this way, the work of the

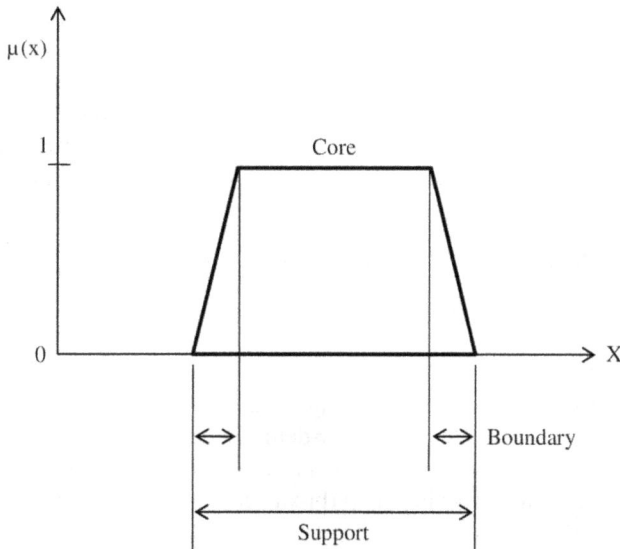

FIGURE 11.1 Graphical representation of features of a membership function.

system designer and the computer is simplified and results in much more precise representation of the manner the system performs in the real world. Generally, fuzzy logic provides a straightforward way to reach a particular conclusion based upon ambiguous, imprecise, vague, noisy or missing input information.

Fuzzy logic could be understood as a superset of conventional (Boolean) logic that has been broadened to figure out the theory of partial truth that is truth values between "completely true" or "completely false".

11.7 FUZZY LOGIC VERSUS BI-VALUED LOGIC

Bi-valued logic can have only two probable values such as yes/no, right/wrong, 0/1, etc., but fuzzy logic can have more than two probable values, so it is multi-valued. It can have many comparative values like yes, no, not so much, a little bit, etc.

11.8 LINGUISTIC VARIABLES AND HEDGES

The initiative of linguistic variables lies at the core of fuzzy set theory. A linguistic variable is regarded as a fuzzy variable. For illustration, let us consider a statement, "John is tall", which means that the linguistic variable (John) acquires the linguistic value (tall). A linguistic variable holds with itself the theory of fuzzy set qualifiers, known as hedges. Hedges are defined as some terms that help in modifying the shape of fuzzy sets. Adverbs such as *more or less, very, somewhat, quite* and *slightly* are included in it.

11.9 FUZZY LOGIC CONTROL SYSTEM

There are two inputs on which the fuzzy logic control system works. The first and second inputs are error and change in error, respectively. The error is acquired by the comparison of input and output signal. If the error obtained is verified with time, then this is known as change in error. There are three main constituents of the fuzzy logic control system: fuzzification, inference mechanism and defuzzification. After the inputs are supplied to the fuzzy logic controller, it decides the output of this controller with the help of fuzzy rules, initially defined by the designer.

11.10 CONSTITUENTS OF FUZZY LOGIC CONTROL SYSTEM

The fuzzy logic control system comprises three main constituents,

Fuzzification: This constituent comprises two more constituents: (i) membership function, and (ii) labels. In this, the input variable is translated into the fuzzy membership function by following a user-defined chart—for instance, motor speed is too low or temperature is too cold—and is allocated grade values starting from 0 to 1. The procedure of converting crisp (bi-valued) input values into linguistic ones is known as fuzzification. There are many types of shapes like Z, A, π, and S which can be utilized by membership functions.

Steps for Fuzzification:

First step: A fuzzy set is developed by converting input values into linguistic concepts.

Second step: To find out the degree of membership function, membership functions are applied to the dimensions.

Inference Mechanism: This constituent comprises fuzzy rules initially defined by the fuzzy controller designer. By using these rules, the output of the fuzzy logic controller is decided by the controller designer. This is the key intelligent part of this system.

Defuzzification: After investigating the fuzzy rules, fuzzy data values are translated into real-life ones by this constituent, but these real-life data values are dependent on the process of defuzzification. Defuzzification translates the fuzzy values into crisp (bi-valued) values. Various procedures such as the centroid method, max-membership method and weighted average method are employed for the process of defuzzification. Each procedure has its own benefits and limitations. These types of procedures are defined by the controller designer.

Methods of Defuzzification:

   i. *Centroid method:* In this method, the center point of the intended fuzzy region is found by computing the weighted average of the output fuzzy region.
   ii. *Max-membership method:* It selects the elements which have maximum value.
   iii. *Weighted average method:* A weight is allocated by a respective maximum membership value to each membership function in the output.

11.11   ADVANTAGES OF FUZZY-BASED SYSTEMS

   i. They are very much comparable to human reasoning.
   ii. Outcomes are obtained with higher precision.
   iii. Simple and straightforward mathematical models are used for cracking real-world problems, whether linear or non-linear.
   iv. They are used in quick operations and to control decisions.
   v. They are very helpful in rule-based modeling and in membership dimensions.

11.12   DISADVANTAGES OF FUZZY-BASED SYSTEMS

   i. Speed is low and requires a long time for processing.

ii. They produce an insufficient real-time response.

iii. User needs to engage a significant quantity of data to get more precise results, which also adds to more rules for reasoning.

iv. They do not use the feedback of the system.

11.13 SOME APPLICATIONS OF FUZZY-BASED SYSTEMS

i. Face recognition.

ii. Handwritten digit/letter recognition.

iii. Automobiles and other vehicle sub-systems: used to manage the velocity of vehicles, in anti-braking systems.

iv. Examine the shape of chromosomes.

v. Temperature regulators: refrigerators, air conditioners.

vi. Socio-economic development.

vii. Cameras: for auto-focus.

viii. X-ray images.

11.14 FUZZY TIME SERIES (FTS) AND ITS APPLICATIONS

Forecasting science plays an important role in research related to economics and finance. It is the most contributed section in such research. Though it is developed on econometric perception, a fresh era in the field of forecasting has started due to recent improvements on fuzzy expanded time series analysis. Rather than the conventional econometrics, forecasting problems can be solved by using FTS, which provides a computer-supported and flexible system. The concept of fuzzy in time series analysis was first advocated by Song and Chissom (1993a), and a time variant and time invariant FTS algorithm was also proposed. Traditional time series models like ARMA, ARIMA and SARIMA are incapable of dealing with problems where the historical data are available in the structure of linguistic constructs in spite of being in the form of numerical values. Various new methodologies have been proposed for dealing with forecasting problems by means of FTS. The major distinction between FTS and traditional time series is that the observations of the FTS are in the form of fuzzy sets while those of traditional time series are in real number form. According to fuzzy modeling, the components of the time series model are regarded as fuzzy sets, and a lot of techniques have been proposed by various researchers. Three groups of FTS models are taken by Yarushkina et al. (2012): first group—a regression model, based upon a fuzzy regression coefficient (Diamond, 1988; Tanaka, 1982), second group—a Box Jenkins model, supported by a fuzzy autocorrelation coefficient (Khashei, 2009; Tsenga, 2000), third group—a fuzzy reasoning model based upon an IF-THEN rule along with the use of an FTS model (Song & Chissom 1993b).

While forecasting the FTS, the main problem is precision in forecasts. Various researchers, (Chen & Hwang, 2000; Huarng, 2001; Hwang et al., 1998; Sullivan & Woodall, 1994) tried to improve the estimates by following low mean square error (MSE) by working on a variety of models for FTS forecasting. A method supported by higher order of FTS was proposed by Chen (2002). The fuzzy logical relations of higher orders—second, third, fourth and fifth—were used for forecasting. On further study of the data, the third order model was found to provide better forecasts on the basis of MSE so obtained. These types of methodologies require more time and a large amount of computational work. To conquer the limitations of previous approaches, a new method of re-splitting the partitions of the universe of discourse and identified fuzzy sets was introduced by Chen and Hsu (2004). A higher order model based upon heuristic function was applied on Taiwan Futures Exchange (TAIFEX) to conquer the uncertainty in trend (Own & Yu, 2005).

Singh (2007) provided simple algorithms of linear order by using a difference operator, and the values attained were utilized in fuzzy rulemaking. This method reduces the time taken by maxima-minima operations and the process of defuzzification. For checking its aptness, the proposed method was implemented on rice production data taken from farms of G. B. P. U. A. & T., Pantnagar. This model was based on first order and gave better results in terms of low MSE. Singh (2009) proposed a method which could be applied when the time series is non-stationary and could not be enlightened by time invariant FTS. This methodology was applied on mustard production data taken at G. B. P. U. A. & T., Pantnagar. To predict the closing price of India's stock market, Badge and Srivastava (2010) used a different approach for FTS by developing technical indicators as input variables.

Recently, Jain et al. (2018b) recommended a computational method for making intervals in FTS forecasting based on the concept of central tendency—mean and variance. This method, accompanied by Gaussian membership, overpowered other methods for FTS. A particle swarm optimization (PSO) fuzzy method was introduced by Jain et al. (2018a) to predict the water table elevation fluctuation. Bose and Mali (2019) summarized the contribution in the area of FTS forecasting during the past 25 years. Goyal and Bisht (2020a) presented a method for FTS forecasting based on an enhanced method for construction of rule base and modified weighted defuzzification. The proposed methods proved their applicability under certain performance evolution criteria. They also proposed a new FTS model based on strong α-cut associated membership (Goyal & Bisht, 2020b).

11.15   FUZZY TIME SERIES (FTS) MODEL

The major distinction between the time series analysis and other statistical analysis is the significance of the order. The observations are statistically independent in other problems but the consecutive observations are dependent in time series analysis. Time series analysis consists of procedures to investigate time series data with the aim of extracting statistics and characteristics related to data which have usual chronological ordering. So, this model is for prediction of future values based upon formerly observed values. The primary function of time series analysis is to recognize the aspect so that the course of future events could be more accurately guessed. In this way, the future performance of the process is controlled by the generating mechanism.

Definition 1: Let a subset of real numbers, $Y(t)$, $(t = 1,2,3, ...)$, be the universe of discourse through which the fuzzy sets $f_i(t)$ are described. Let

F(t) be the collection of $f_1(t)$, $f_2(t)$, $f_3(t)$,..., then F(t) is termed as an FTS defined upon Y(t).

Defintion 2: Assume that F(t) is computed by F(t–1) only. Also, F(t) can be written as, F(t) = F(t–1) R (t–1), where R(t–1) is known as a fuzzy relationship for F(t–1) and F(t). For some t, if R (t–1, t) is not dependent upon t, then F(t) is regarded as a time invariant FTS. If not, then F(t) is time variant FTS.

FTS is considered to be a fuzzy variable together with a related membership function. A procedure for fitting the FTS model has been put forwarded by Song and Chissom (1993). This procedure consists of a fuzzy set, F, described on a universe of discourse U, $\mu F(x)$: U $\rightarrow$[0, 1] and membership function μ. This is a mapping from U to the unit interval [0, 1], and $\mu F(x)$ symbolizes the level to which F is contained in x. This concept permits an element belonging to U to provide partial membership to a particular fuzzy set and other fuzzy sets, also. Time variant and time invariant models were applied by Song and Chissom to predict the employment at the University of Alabama. This model consists of the following steps:

1. Defining the universe of discourse and the intervals

Universe of discourse, U, is based upon the ranges which are present in the available past time series data and is obtained by the following rule: U = $[B_{max} - B_1, B_{min} - B_2]$, where B_1 and B_2 are the two appropriate positive numbers.

1. Dividing the intervals
   The universe of discourse is divided into identical length of intervals i.e. u_1, u_2, u_3, ... u_n. The number of intervals, n, should be in the unity with respect to the number of linguistic variable (fuzzy sets), A_1, A_2, A_3, ...A_n are to be concerned. After that, the central point of the intervals is to be found.
2. Classifying the fuzzy set
   The fuzzy set is classified into linguistic variables, A_1 = very good, A_2 = good, A_3 = poor, A_4 = very poor, and so on. A fuzzy set is always present with the help of the significance of each fuzzy variable. Each linguistic value corresponds to a fuzzy variable. $A_i = (\mu A_i (u_i)/u_i)$, $u_i \in$ U and $\mu A_i(u_i) \in [0,1]$ is regarded as a fuzzy set if the value of variable U is recognized as the central point of the corresponding interval of fuzzy set.
3. Fuzzifying the data
   In this step, the historical data is fuzzified, and the fuzzy logical relationships are set up by the subsequent rule: let A_i be the fuzzy value for n and A_j be the fuzzy value for n + 1, then the fuzzy logical relationship is represented as $A_i \rightarrow A_j$, where A_i is the present stage and A_j is the next stage.
4. Defuzzifying the forecasting outcomes
   i. To find the average of the data
   ii. To find the sum of the data values and divide the obtained sum by the number of data values

iii. To calculate the absolute value of the difference between each data value and the average

iv. To compute the total or absolute value of the difference.

11.16 CONCLUSION

Fuzzy logic is an indispensable element of technology. It provides services which can be applied to solve a variety of real-world problems. The problem-solving technique is greatly flexible and transparent to scale and decreases the conditions required according to the application needs. FTS methodologies were built up because the traditional statistical time series methods were not able to examine time series efficiently with a small quantity of data. Apart from this requirement, FTS methods also provide an instrument to deal with situations where historical data are in the form of linguistic values. The forecasting science has achieved its limits in theoretical framework but the FTS method is a fast-evolving field and advancing in its consistency and precision. As it is a computer intelligence-based technique, the FTS method has come as an alternate to the traditional methods due to its capacity of unaided-automatic structure. A debate is present on the topic of clashes between the traditional econometrics and the FTS methodology. The conventional econometrics has added several theoretical derivations to the existing literature. But, on the other hand, the existing research on FTS generally pays no attention to well-known laws and principles, and therefore its practical value is doubtful. Since the performance of already-existing forecasting methods is well acknowledged, FTS models are essentially supposed to guarantee supremacy to them. To increase the practical value of FTS, it is necessary to talk about the theoretical bases of econometrics and implement them with FTS models. Also, additional advancement on automatic control and computation should be appreciatively recognized by the wider society, and generalization should be provided.

REFERENCES

Badge, J., & Srivastava, N. (2010). Asset value estimation through technical indicators and fuzzy time series model. *International Journal of Computational and Applied Mathematics, 5,* 437–446.

Bajestani, N. S., & Zare, A. (2011). Forecasting TAIEX using improved type 2 fuzzy time series. *Expert Systems with Applications, 38*(5), 5816–5821.

Bose, M., & Mali, K. (2019). Designing fuzzy time series forecasting models: A survey. *International Journal of Approximate Reasoning, 111,* 78–99.

Chatfield, C. (1993). Calculating interval forecasts. *Journal of Business & Economic Statistics, 11,* 121–144.

Chen, S. M. (2002). Forecasting enrollments based on high-order fuzzy time series. *Cybernetics and Systems: An International Journal, 33,* 1–16.

Chen, S. M., & Hsu, C. C. (2004). A new method to forecast enrollments using fuzzy time series. *International Journal of Applied Sciences and Engineering, 2,* 3234–3244.

Chen, S. M., & Hwang, J. R. (2000). Temperature prediction using fuzzy time series. *IEEE Transaction on Systems, Man & Cybernetics, 30,* 263–275.

Diamond, P. (1988). Fuzzy least squares. *Information Sciences*, *46*(3), 141–157.

Dubois, D., & Prade, H. (1991). Fuzzy sets in approximate reasoning, Part 1: Inference with possibility distributions. *Fuzzy Sets and Systems*, *40*, 143–202.

Gneiting, T., & Katzfuss, M. (2014). Probabilistic forecasting. *Annual Review of Statistics and Its Application*, *1*, 125–151.

Goyal, G., & Bisht, D. C. S. (2020a). Fuzzy time series forecast with enhanced trends and weighted defuzzification. *Mathematics in Engineering, Science & Aerospace (MESA)*, *11*(1), 91–102.

Goyal, G., & Bisht, D. C. S. (2020b). Strong α-cut and associated membership based modeling for fuzzy time series. *International Journal of Modeling, Simulation, and Scientific Computing*, 12(1), 2050067:1–2050067:20.

Huarng, K. (2001). Heuristic models of fuzzy time series for forecasting. *Fuzzy Sets and Systems*, *123*, 369–383.

Huarng, K., & Yu, H. K. (2005). A type 2 fuzzy time series model for stock index forecasting. *Physica A: Statistical Mechanics and its Applications*, *353*, 445–462.

Hwang, J. R., Chen, S. M., & Lee, C. H. (1998). Handling forecasting problem using fuzzy time series. *Fuzzy Sets and Systems*, *100*, 217–228.

Jain, S., Bisht, D. C. S., & Mathpal, P. C. (2018a). Particle swarm optimized fuzzy method for prediction of water table elevation fluctuation. *International Journal of Data Analysis Techniques and Strategies*, *10*(2), 99–110.

Jain, S., Mathpal, P. C., Bisht, D. C. S., & Singh, P. (2018b). A unique computational method for constructing intervals in fuzzy time series forecasting. *Cybernetics and Information Technologies*, *18*(1), 3–10.

Khashei, M. (2009). Improvement of auto-regressive integrated moving average models using Fuzzy logic and artificial neural networks. *Neurocomputing*, *72*(4), 956–967.

Own, C. M., & Yu, P. T. (2005). Forecasting fuzzy time series on a heuristic high-order model. *Cybernatics and Systems: An International Journal*, *36*, 705–717.

Sadaei, H. J., Enayatifar, R., Guimaraes, F. G., Mahmud, M., & Alzamil, Z. A. (2016). Combining ARFIMA models and fuzzy time series for the forecast of long memory time series. *Neurocomputing*, *175*, 782–796.

Singh, P. (2017). A brief review of modeling approaches based on fuzzy time series. *International Journal of Machine Learning and Cybernetics*, *8*, 397–420.

Singh, S. R. (2009). A computational method of forecasting based on high order fuzzy time series. *Expert Systems and Applications*, *36*, 10551–10559.

Singh, S. R. (2007). A simple method of forecasting based on fuzzy time series. *Applied Mathematics and Computation*, *186*, 330–339.

Song, Q., & Chissom, B. H. (1993a). Fuzzy forecasting enrolments with fuzzy time series-Part 1. *Fuzzy Sets and Systems*, *54*(1), 1–9.

Song, Q., & Chissom, B. H. (1993b). Fuzzy time series and its models. *Fuzzy Sets and Systems*, *54*, 269–277.

Sullivan, J., & Woodall W. H. (1994). A comparison of fuzzy forecasting and Markov modeling. *Fuzzy Sets and Systems*, *64*(3), 279–293.

Tanaka, H. (1982). Linear regression analysis with fuzzy model. *IEEE Transactions on Systems, Man and Cybernetics*, *12*(6), 903–907.

Tsenga, F. M. (2000). Fuzzy ARIMA model for forecasting the foreign exchange market. *Fuzzy Sets and Systems*, *118*, 9–19.

Yarushkina, N., Afanasieva, T., & Perfilieva, I. (2012). Computational intelligence in the analysis of time series. *"FORUM": INFRA-M*, *160*, 160.

Zadeh, L. A. (1965). Fuzzy sets. *Inform and Control*, *8*, 338–353.

12 Artificial Neural Networks (ANNs) and their Application in Soil and Water Resources Engineering

M. Mohan Raju[1], Dinesh C. S. Bisht[2], A. Naresh[3], Harish Gupta[4], and M. Gopal Naik[5]

[1]Nagarjunasagar Project, Irrigation & CAD (Projects Wing) Department, Government of Telangana State, Hill Colony-508202, India

[2]Department of Mathematics, Jaypee Institute of Information Technology, Noida-201304, India

[3]Department of Civil Engineering, Osmania University, Hyderabad-500007, India

[4]Department of Civil Engineering, Osmania University, Hyderabad-500007, India

[5]Department of Civil Engineering, Osmania University, Hyderabad-500007, India

12.1 INTRODUCTION

Machine learning (ML) is an interdisciplinary area of research involving applied conceptual understanding of diversified fields of artificial intelligence (AI), information technology, statistics, control theory, cognitive science, biology, philosophy, etc. Artificial neural networks (ANNs), fuzzy logic, reinforcement learning (RL) and related methods are generally employed in ML when developing computer programs for decision-making capability by learning from data or experience/training. The learning process is also known as training that the neural network itself organizes to develop an internal set of features to classify the data or information. ANN is a soft computing method that represents highly idealized mathematical models of complex systems understanding inspired by the human brain functioning studies and the nervous system in biological organisms and having an ability to learn from an experience. ANN is not a conventional computer program but describes or presents with examples of the pattern recognition,

DOI: 10.1201/9781003102281-12

concepts and observations or other data supposed to be learnt. ANN is an efficient soft computing method for complex situations due to its massively parallel processing architecture for high-speed huge data processing.

ANNs have been around since the 1940s. Later, various algorithms evolved to overcome the limitations of the early networks in real-time applications of ANNs. The algorithms developed in the recent past have a main objective of input and output mapping and have been proven to yield comparatively better results over conventional mapping during the applications of (1) poorly described complex systems; (2) specified problems like pattern recognition, generalization, diagnosis and situations to deal with noise; and (3) ambiguous natured or incomplete input situations. ANNs have good pattern recognition; hence, they do not need to develop a relation (linear, power, polynomial, etc.) between independent and dependent parameter directly. Further, many simple and lucid characteristics of ANN methodology can be seen in problem-solving instances: (1) no prior knowledge requirement for a neural network when working with an underlying process; (2) recognition of the complex relationships that exist among various aspects of the investigated process that may not be defined during execution; (3) a statistical model analysis or a standardized optimization yields a final result for a specified problem at the end of its run, but in the case of a neural network an optimal or suboptimal convergence proceeds to the solution; (4) no prior structure of solution and constraints are defined or enforced strictly in the development of ANN methodology. These characteristics of ANNs render a possible and very suitable solution to handle the many hydrologic phenomena like rainfall-runoff, runoff-sediment and stage-discharge modeling problems.

ANN is an architectural network of interconnecting artificial neurons or nodes, and each individual neuron will have a local input/output (I/O) functional computation. The neuron's output is usually determined by it's input/output characteristics, interconnections with other neurons/units and required external inputs possible, if any. The required design alteration or modification of the network is also conveniently allowed during the training process; however, the network progresses towards complete functionality in general by means of one or more methods of training. Approximately 60 year ago, the methodology development of ANN began (McCulloch & Pitts, 1943), based on the inspiration of a neuron in the biological system of human brain functioning. The efficient pattern recognition nature in human brain processing inspired the development of ANNs. ANNs are the soft computing mathematical models that mimic the function of the human brain and nervous system. The developmental progress has the following rules:

   i. Node is an information processing element called a unit or neuron, and the processing occurs at many single nodes in the network.
  ii. Signals are transmitted or passed among the nodes through interconnected links.
 iii. Each connection link has connection strength due to an associated weight of the connection link.
 iv. To determine the output signal, a non-linear transformation or an activation function is applied to the net input.

12.2   BIOLOGICAL BASIS OF ANNs

A biological neuron system, also known as a nerve cell system in the human body, is a complex system of biochemical and electrical signal processing. The main cell body of a biological neuron is called a soma and consists of a nucleus that regulates and controls the cell activity like growth and metabolism. Signals are received by the nerve fibers, called dendrites, to transmit to the soma, i.e., the fundamental processing unit and its outer boundary covered with cell membrane. Further, the cell interior and outside is filled with intracellular fluid and extracellular fluid, respectively. The axon is a cellular connection extending from the cell body that connects many other neurons through its stands and sub-stand branches at the synapses or synaptic junction/terminal through which the electrical signal is transmitted from one neuron to the other when a neuron fires. The axon typically leads to many synapses of other surrounding neurons. The signal transmission is a complex chemical process at the synaptic junction where specific transmission substances are released from the sending side of the junction. The cell is said to have fired when short pulses of electrical activity are generated due to a threshold potential. Synapses are exclusively operated by chemical mechanism, and the electrical activity is confined to the interior of the neuron. This generated potential will then be transmitted by axon to the other neurons/cells. The dendrites are the main receptors of the signals from other neurons to transmit the information or signal to the soma and finally to transmit it to the nucleus for processing the data. The information from sensory organs like eyes, ears, the nose or muscles is usually received by a receptor neuron, which is a third type of neuron that exists in the system. We suggest you go through the literature to understand the development of the ANN system on the basis of biological neurons. The current chapter strictly followed the rules and guidelines of ASCE (2000a; 2000b) in the research; hence, it is requested that the reader can go through the same for clear understanding of the discussed applications.

12.3   MODEL OF AN ARTIFICIAL NEURAL NETWORK

An individual signal processed at a single artificial neuron is shown in Figure 12.1. The input signals or data are fed to the single neuron network. Later, the individual signals are multiplied by their respective connecting strengths or weights, through which they are transmitted towards the processing neuron and summed up at this juncture further to produce the final output. The net input is then transformed through the transformation function known as the activation function
where

x_1, x_2, x_3, x_4, x_5,......x_n are the inputs and their respective weights are w_1, w_2, w_3, w_4, w_5, ...w_n.

The transmission of net input fed to the node is given by

$$net = \sum_{i=1}^{n} x_i w_i \qquad (12.1)$$

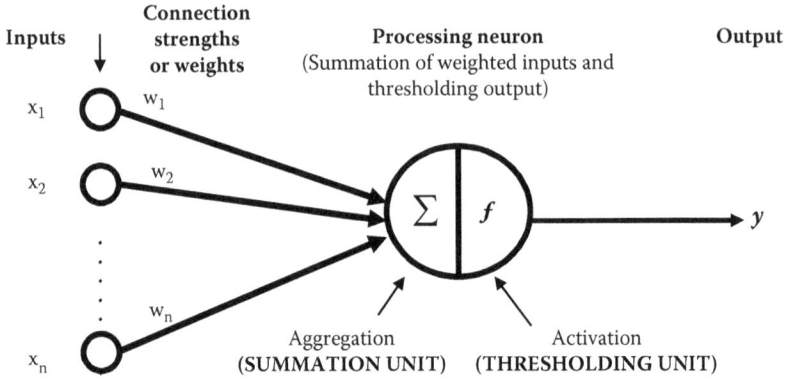

FIGURE 12.1   A single processing unit (artificial neuron).

Now, the activation function $f(.)$ is used to pass the net input through it to compute the final output "y".

$$y = f(net) \tag{12.2}$$

The most commonly used nonlinear activation function used in general in solving ANN problems is sigmoidal transfer and is given by $y = f(net) = \frac{1}{1+e^{-net}}$.

i. The topology of ANN

The topology of ANN comprehends the connection infrastructure of an ANN. Connection is the bridge between two neurons to transport the data and is also classified into inhibitory and excitatory. To stop a neuron to react (negative term in a sum), inhibitory connections are employed, and excitatory connections are employed for firing of the neuron (positive sum). Delay connections are another type of connection generally applied in the data transmission, while a lag time is necessarily included, which is essential in time series researches like flood routing, overflow over a weir, etc., and their analyses (Day & Davenport, 1993). Layers are classified into input layers, hidden layers and output layers. Input layers receive the data from outside world and transmit them to hidden layers, where the data will undergo various operations and internal optimizations before transmitting to the output layers. Finally, the estimated data will be presented to the outside world through the output layers. During the transmission of the data from the input to the output side, the data will pass more than once through all the available neurons or some of the available neurons in the network, as the current existence condition of the network is dependent on the previous status of the network, which implies that the same input vectors result in various and different output vectors. Hence, the feedback networks are referred to as dynamic non-linear systems in the optimization process of the output (i.e., during network data transmission from input to the output and vice versa), where convergence and stability characterize the performance of a feedback network (Kohonen, 1998).

ii. **Learning algorithms**

The learning process in the ANN system involves adjustment of weights in the system by means of example sets used for training the network and learning algorithms. In general, the learning or training process is carried out in neural network methodology in three modes: supervised learning, unsupervised learning and batch learning. In supervised training, a desired response will be provided immediately when the input is given in the network and an indication is given that the network performs. When the training is unsupervised, the desired response is unknown, and the weights are adjusted based on the responses with limited knowledge or no knowledge of the inputs. When the total training data samples used to create one batch are fed to the network for learning, the learning algorithm is called batch gradient descent.

iii. **Structure**

In order to learn properly, an ANN must be trained or provided by ample amount of examples. The qualitative and correct suitable data for training should be chosen and pre-processed accordingly. It is crucial for a network to perform well; otherwise, it results in poor learning and unreliable network performance (Haykin, 1994). The most important pre-processing for input data is normalization, i.e., filtering and scaling. To reduce the influence of outliers present in the data series/time series, the data must be normalized and may be brought into a uniform range [0 1] as a common practice. Low frequency signals and high frequency signals may influence the main signal to alter its regular state or path when working with time series data; hence, the data needs to be filtered for smooth transmission in the network (Karunanithi et al., 1994). However, ANNs themselves have a tendency of filtering time series data during data processing while transmitting the signal to the network towards the output.

iv. **Architecture**

ANN architecture describes the information about the neurons present in the network and how they are interconnected with each other, along with the structural layout of the hidden layers and clusters in the network. The ANN architecture is often modified or changed to improve ANN performance, for example, an increase or decrease in the number of neurons/nodes from the input side or an increase or decrease in the number of hidden layers and nodes in the layers (Refenes & Vithlani, 1991). In addition to these modifications, modularization can be applied to improve ANN performance (Nadi, 1991). For example, in simulating the actual outflow of a sewer system, a large quantity of observed time series data that is very complex in nature will be restructured into many abstract parameters and will be fed to another ANN for final output by employing a self-organizing neural network.

v. **Multilayered feed-forward ANN**

The unique feature and characteristic nature of multilayered feed-forward ANN is that it can map any complexity in the time series data (Zurada, 1992). The multi-layered feed-forward ANN architecture depicts information about the network such as how the neurons are grouped in the network and the input layer, the output layer along with the number of intermediate layers known as hidden layers and their associated neurons present in the network. The network also gives information about interconnecting strengths among the neurons that arise due to interconnections between layers, known as weights, associated with the input signal in the transmission from the input layer to the output side. The interconnected number of neurons or nodes are considered in a series in the following manner: $i = 1, 2, 3......ni$ for the input layer, $j = 1, 2, 3 ....nh$ for the hidden layer and $k = 1, 2, 3.......no$ for the output layer (Figure 12.2).

vi. **BPNN back propagation neural network**

Extensively used by nearly 80% of today's water resources applications is the back propagation (BP) neural network in the areas of simulation, modeling, prediction and the uniqueness of the network. During the learning process, the error is propagated back through the network from the output side to the input side to modify the weights (Figure 12.3), due to which the network is assigned the name BPNN. Multilayer feed-forward ANNs are systematically trained based on the error-correction rule by employing the BP algorithm among the layers of the network where the connecting strengths or weights are updated by reduction of errors between observed and estimated values (Kothari and Agyepong 1996).

Figure 12.3 depicts the directions of input signal transmission towards the output world for estimation or prediction in a multi-layered ANN and the error BP towards the input side. Normally, learning or training is an iterative process in BP algorithm

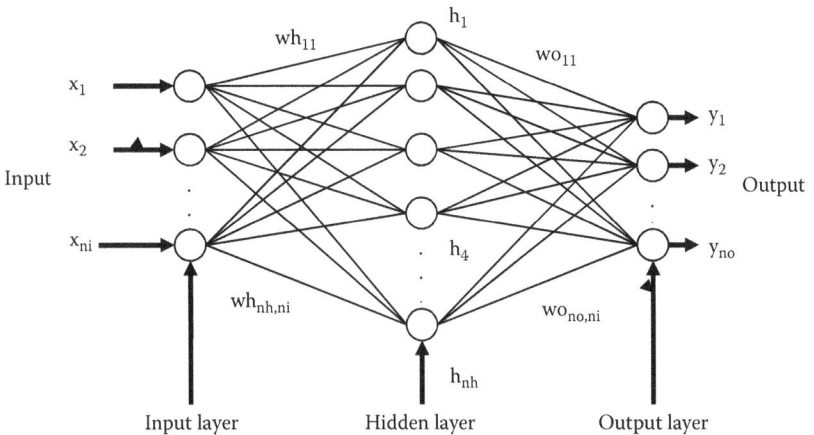

FIGURE 12.2   Architecture of multilayered feed-forward ANN configuration.

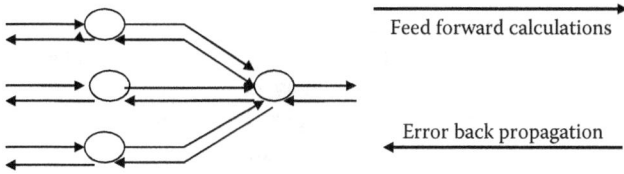

FIGURE 12.3   Directions of signal transmission in multi-layered ANN.

application, where the errors are propagated back and the randomly initialized weight parameters are updated in every iteration through feed-forward calculations.

vii. Learning factors of back propagation

The convergence of BPNN is based on important learning factors, which is a remarkable problem of the BP algorithm to be optimized. The main learning factors include the network architecture, learning rate, initial weights and the nature of the training set (Karnin, 1990).

a. Initial weights

In multi-layered feed-forward neural networks, the initial weights are typically assigned at random (between –1.0 and 1.0, or –0.5 to +0.5), which affects the final output of the network. The initial weights should not be large from the beginning of the iterative process, which may saturate the sigmoid activation function and lead to stock at local minima. Further, the equal weights in the iteration process of the network also leads to improper training when a solution with different weights is needed in the presentation. Even the range of initialized weights should not be too small as it results in very slow learning. Hence, the saturation is avoided by selecting the uniformly distributed synaptic weights' initial values inside a small range of values.

b. Frequency of weight updates

  i. Updating of weights after completion of each sample presentation to the network i.e. "per-pattern" learning approach.
  ii. Updating of weights after complete or total samples presented to the network i.e. "per-epoch" (or "batch-mode") learning approach. An *epoch* consists of the entire set of training samples.

c. Learning rate (η)

Learning rate determines the updating or changing of weights during the iterative process. Weight vector updates or modifications in BP are proportional to the negative gradient of the error that occurs during iterations. The magnitude of weight alteration is dependent on the suitable selected value of "η", and a comparatively

bigger value of "η" selection results in rapid learning; thereby, unwanted oscillations may be observed in the weights. As far as the selection of lower values for "η"
is considered slow learning, the process may be seen during training. Hence, the
selection of an appropriate value for "η" depends on the applications under study as
it is a distinct operation followed in gradient descent methods. In many water resources engineering studies and literature presented in this chapter, the value for "η"
is assigned in the range between 0.1 and 0.9.

d. Momentum rate (α)

In the gradient descent method, most commonly the momentum rate is included or
applied to avoid the danger of instability and to increase the learning rate (Sato,
1991). The most common practice of utilizing a value for "α" is 0.9, and its range
falls in the closed interval from 0 to 1, i.e., [0,1]. A value of 0.9 for "α" was chosen
in the ANN application of research topics discussed below. An inertia or momentum component is given to the weights to change the downhill force (generated
on an average) direction.

e. Normalization of the data

Normalization or standardization is the process of bringing the input data variable
by scaling range from 0 to 1, because normalization brings the noise effects to a
mean level and smooths the solution space.

f. Assessment on the count of hidden layers and nodes

The number of hidden layers and nodes are determined experimentally, i.e., most of
the times, it is a hit-and-trial method. In practical applications, it is seen to have a
lesser number of hidden layers than the number of input parameters that is fed to the
network as inputs. Additionally, when a network requires more neurons for better
convergence in the network performance, gradually the number of neurons may be
enhanced until the optimized convergence is attained. Here, in the current research
examples, for better convergence experience, inclusion of one hidden and two hidden
layers has been employed, and in the case of hidden nodes in the hidden layers, the
number was equal to or double the number of nodes present in the input layer.

g. Data training and generalization

The data provided for training the network should be sufficient enough for good
generalization capability of the network in mapping the input and output, and as such,
there is no specific rule for deciding the magnitude or quantity of the data to be submitted to the network for training. The working data are usually allocated to some for
training and testing. Sometimes a third part of data is also considered for the validation
phase. The proper training of the network ultimately leads to the assignment of

optimized connecting strengths that estimates the outputs significantly close to the observed values.

viii. **Calculations of feed-forward network configuration**

In a feed-forward multi-layered neural network, input signals are fed to the input layer, and they pass through hidden layers, then through the output layer, to yield a final output. When the signals are passing from one layer to the other, the respective connection strengths or weights are given to them for multiplication and to produce weighted vector inputs. Then, these weighted vectors are aggregated to generate a net input at an intermediate neuron. Subsequently, this generated net input passes through an activation function or the thresholding unit to transmit its net signal to the next layer and so forth to result in a final output.

Net input received by the hidden layer at the jth node is estimated as follows:

$$neth_j = \sum_{i=1}^{ni} wh_{ji} x_i \ldots \tag{12.3}$$

where ni = no. of neurons in the input layer and wh_{ji} = connecting strength between i^{th} node of the input layer and j^{th} node of the hidden layer. The output of jth node of the hidden layer h_j is given by

$$h_j = f\left(neth_j\right) \tag{12.4}$$

$f(.)$ is the usual sigmoidal transfer/activation function, and operated as follows

$$h_j = \frac{1}{1 + e^{\left(-neth_j\right)}} \tag{12.5}$$

Further, the net input received by the output layer at k^{th} node is estimated as follows:

$$nety_k = \sum_{j=1}^{nh} wo_{kj} h_j \tag{12.6}$$

where nh = no. of neurons in the hidden layer and wo_{kj} = connection strength between j^{th} node of the hidden layer and k^{th} node of the output layer. The output of kth node of the output layer can be written as

$$y_k = f(nety_k) \tag{12.7}$$

The operating sigmoidal transfer function is then

$$y_k = \frac{1}{1 + \exp(-nety_k)} \qquad (12.8)$$

Now, the error found between the estimated and targeted outputs is back propagated towards the input layers for further modification, as discussed below.

ix. **Error back propagation**

The calculated error at the output side is back propagated to the input side layers through hidden layers to determine the updated or modified weights (Hario & Jokinen, 1991). In a single input-output mapping, the sum square error (E) of the data set can be written as

$$E = \frac{1}{2} \sum_{k=1}^{no} (y_k - t_k)^2 \qquad (12.9)$$

where y_k = estimated output value at the kth node and t_k = target or observed output value at the same node or neuron. The incremental changes of the connecting strengths calculated in subsequent trials are subtracted from their previous strengths or weights and the deviations or errors are then propagated back to the input layers to minimize the error function in the next trial as follows.

$$wo_{kj}^{new} = wo_{kj}^{old} - \varDelta wo_{kj} \qquad (12.10)$$

$$wh_{ji}^{new} = wh_{ji}^{old} - \varDelta wh_{ji} \qquad (12.11)$$

$\varDelta wo_{kj}$ and $\varDelta wh_{ji}$ are increments in connecting strengths between the output layer and its corresponding hidden layer and hence,

$$\varDelta wo_{kj} = \eta \left(\frac{\partial E}{\partial wo_{kj}} \right) \qquad (12.12)$$

Applying the chain rule, $\frac{\partial E}{\partial wo_{kj}}$ is re-written as

$$\frac{\partial E}{\partial wo_{kj}} = \frac{\partial E}{\partial y_k} \frac{\partial y_k}{\partial nety_k} \frac{\partial nety_k}{\partial wo_{kj}} \qquad (12.13)$$

Differentiating equation (12.6) with respect to wo_{kj},

$$\frac{\partial nety_k}{\partial wo_{kj}} = h_j \qquad (12.14)$$

Differentiating equation (12.8) w.r.t $nety_k$,

$$\frac{\partial y_k}{\partial nety_k} = y_k\left(1 - y_k\right) \tag{12.15}$$

Differentiating equation (12.9) w.r.t y_k

$$\frac{\partial E}{\partial y_k} = (y_k - t_k) = \delta y_k \tag{12.16}$$

where δy_k = the error of kth output node at the output layer side.

Using equations (12.14) and (12.16) in equation (12.13), then equation (12.12) designated as,

$$\Delta wo_{kj} = \eta \delta y_k \cdot y_k\left(1 - y_k\right) \cdot h_j \tag{12.17}$$

or

$$\Delta wo_{kj} = \eta \Delta y_k h_j \tag{12.18}$$

where,

$$\Delta y_k = \delta y_k \cdot y_k (1 - y_k) \tag{12.19}$$

Δy_k = error of kth output node at the input layer. Likewise, for the intermediate or hidden layers, the incremental changes in the weights are expressed by

$$\Delta wh_{ji} = \eta\left(\frac{\partial E}{\partial wh_{ji}}\right) \tag{12.20}$$

Applying chain rule

$$\frac{\partial E}{\partial wh_{ji}} = \sum_{k=1}^{no} \frac{\partial E}{\partial y_k} \cdot \frac{\partial y_k}{\partial nety_k} \cdot \frac{\partial nety_k}{\partial h_j} \cdot \frac{\partial h_j}{\partial neth_j} \cdot \frac{\partial neth_j}{\partial wh_{ji}} \tag{12.21}$$

Differentiating equation (12.8) w.r.t $\partial nety_k$,

$$\frac{\partial y_k}{\partial nety_k} = y_k\left(1 - y_k\right) \tag{12.22}$$

Differentiating equation (12.6) w.r.t ∂h_j,

$$\frac{\partial nety_k}{\partial h_j} = wo_{kj} \tag{12.23}$$

Differentiating equation (12.5) w.r.t $\partial neth_j$,

$$\frac{\partial h_j}{\partial neth_j} = h_j\left(1 - h_j\right) \tag{12.24}$$

Differentiating equation (12.3) w.r.t ∂wh_{ji},

$$\frac{\partial neth_j}{\partial wh_{ji}} = x_i \tag{12.25}$$

Substituting equations (12.16), (12.22) and (12.25) in equation (12.21), then the equation 12.20 is designated as,

$$\Delta wh_{ji} = \eta\left(\frac{\partial E}{\partial wh_{ji}}\right) = \eta \sum_{k=1}^{no} \delta y_k y_k (1 - y_k) wo_{kj} h_j (1 - h_j) x_i$$
$$\Delta wh_{ji} = \eta \sum_{k=1}^{no} \delta y_k y_k (1 - y_k) wo_{kj} h_j (1 - h_j) x_i \tag{12.26}$$

Using equation (12.19)

$$\Delta wh_{ji} = \eta \sum_{k=1}^{no} \Delta y_k wo_{kj} h_j (1 - h_j) x_i \tag{12.27}$$

where

$$\delta h_j = \sum_{k=1}^{no} \Delta y_k wo_{kj} \tag{12.28}$$

Here, δh_j = error at the output side of the jth hidden node.

$$\Delta h_j = \delta h_j h_j\left(1 - h_j\right) \tag{12.29}$$

Here, Δh_j = error at the input side of the jth hidden node.
    Substituting equation (12.29) in equation (12.27)

$$\Delta wh_{ji} = \eta \Delta h_j x_i \tag{12.30}$$

In the network training process, weights are assigned at random and by using equation (12.17) or (12.18) and equation (12.30), weights or connecting strengths are modified at a later stage through the implementation of the back propagation (error back propagation) algorithm.

x. ANN architecture selection

An ANN architecture design criterion is the most important activity of a multi-layered network. The number of nodes in the input layer should be equal to the number of generalized inputs to be fed to the network, and for the output, the nodes are equal to the desired numbers to be estimated. The selection of the number of hidden layers and nodes associated with each hidden layer is a usual practice of effective trial-and-error procedure through which various combinations of hidden layers and nodes will be trained. A suitable network that yields a minimum or low root mean square error (RMSE) value is selected to present the output of the network, and care must be taken in presenting the optimized and smallest network size possible to analyze the results with less complexity. Akaike's Information Criterion (AIC), a highly efficient and best optimized criterion to select the best network, is the minimum RMSE and smallest possible size or architecture. Moreover, the AIC is usually employed to trade the size of the neural network in an operational way as how well the data is fitted to the network. In turn, the lower the AIC value, the better the network is to sort out or work with.

$$AIC = 2m + z \ln\left(\frac{2\Psi}{z}\right) \qquad (12.31)$$

where m = no. of parameters; z = no. of observations and Ψ = sum-square-error.

12.4 APPLICATIONS

The current sub-topic describes our experience with applications of various ANN methodologies in the fields of simulation and modeling of runoff, sediment, river stage, water table elevation fluctuation and spring discharge modeling, along with the database support of rainfall, evaporation and temperature at some instances. A light blend of fuzzy logic will also be seen in one of the analyses and will provide an idea of the hybrid modeling phenomenon.

i. Runoff-Sediment Modeling

The estimation and assessment of sediment runoff in river systems has great concern in hydrologic and hydraulic engineering systems analysis. The assessment of sediment transportation in river systems plays a crucial role in varied hydro-mechanics problems like the design of channels, reservoirs, dams and other hydraulic structures. The hydrologic variables of a catchment such as precipitation, evaporation, evapotranspiration, infiltration and the groundwater table interact with the hydraulic variables such as riverbed slope, bed forms, flow velocity, stage, discharge and turbulence to result in a complex phenomenon to simulate and model

(Wu et al., 1993). The sediment runoff of the watershed is impacted by rainfall and soil detachment that is ultimately deposited downstream of the catchment.

A sediment discharge relationship of a river was modeled using ANN with the available data of sediment concentration and discharge. Govindaraju and Kavvas (1991) and Raju (2008) selected six gauging stations of the Pranhita river, a sub-basin of the Godavari river in India. A comparative study was carried out between the sediment rating curves (SRCs) and ANN to examine the superiority of ANN over classical SRCs. The current study aimed at developing ANN models for the runoff-sediment process using past runoff and sediment concentration values as inputs with specific timestep or a lag. Feed-forward multilayered ANN technique was employed in the study, which uses the back propagation algorithm for ANN training. The study was carried out for the Pranhita sub-basin that influences the Godavari river system to a great extent as it covers an area of 1,09,100 km^2 (34% drainage area of Godavari basin). The study was framed to analyze the following objectives:

  i. *Developing the runoff-sediment ANN models and SRCs for the Pranhita sub-basin of the Godavari river system.*
 ii. *To validate the developed models.*
iii. *To evaluate the performance of the formulated models for the Pranhita sub-basin under study.*

a. **Runoff-sediment models and input parameters**

The runoff and sediment concentration data procured for the study in a daily interval basis at the six hydrological gauging stations—Tekra (Pranhita river), Penganga Bridge (Penganga river), Bamni (Wardha river), Hivra (Wardha river), Ashti (Wainganga river) and Pauni (Wainganga river)—for four years from June 2000 to May 2004, constituting a total of 1,460/1,462 patterns. Out of this, 730/731 patterns were used for training, 365/366 patterns for testing and 365/366 patterns for validation. The output or the estimation is the sediment concentration designated at timestep t, i.e., C_t and has been cited in literature by many authors (Agarwal et al., 2006; Anmala et al., 2000; ASCE, 2000a; ASCE, 2000b, Tayfur & Guldal., 2005). The current sediment concentration can be mapped better by considering, in addition to the current value of discharge, the sediment and discharge at the previous times (Jain, 2008). Therefore, in addition to Q_t, i.e., discharge at timestep t, other variables such as Q_{t-1}, Q_{t-2}, and C_{t-1}, C_{t-2}, were also considered in the input (Cigizoglu, 2002; Raghuvanshi et al., 2006). The input data considered to train the ANN with various combinations for the current study are shown in detail in Table 12.1. However, in the design of SRC analysis, the input-output variables used for ANN-1 were only used in modeling, as the variables are prime and fundamental variables in the development of the models to start with (Naresh et al., 2019). The development of the ten models executed in the following fashion are shown in Table 12.1. Various numbers of trials were made, with one and two hidden layer methodology for a particular model, and the best representative models were

TABLE 12.1

Runoff-Sediment Models Developed for the Study

Model	Hidden Layers Applied	Output	Inputs
ANN – 1	One	C_t	Q_t
ANN – 2	One	C_t	Q_t, Q_{t-1}, C_{t-1}
ANN – 3	One	C_t	$Q_t, Q_{t-1}, Q_{t-2}, C_{t-1}, C_{t-2}$
ANN – 4	One	C_t	$Q_t, Q_{t-1}, Q_{t-2}, Q_{t-3}, C_{t-1}, C_{t-2}, C_{t-3},$
ANN – 5	One	C_t	$Q_t, Q_{t-1}, Q_{t-2}, Q_{t-3}, Q_{t-4}, C_{t-1}, C_{t-2}, C_{t-3}, C_{t-4}$
ANN – 6	Two	C_t	Q_t
ANN – 7	Two	C_t	Q_t, Q_{t-1}, C_{t-1}
ANN – 8	Two	C_t	$Q_t, Q_{t-1}, Q_{t-2}, C_{t-1}, C_{t-2}$
ANN – 9	Two	C_t	$Q_t, Q_{t-1}, Q_{t-2}, Q_{t-3}, C_{t-1}, C_{t-2}, C_{t-3},$
ANN – 10	Two	C_t	$Q_t, Q_{t-1}, Q_{t-2}, Q_{t-3}, Q_{t-4}, C_{t-1}, C_{t-2}, C_{t-3}, C_{t-4}$

Where Q = discharge, C = sediment concentration

selected for presentation of multilayer neural network. Ultimately a best-suited model was drawn from two methodologies to represent the given gauging station (Figure 12.4).

b. Rating curves for sediment estimation

SRCs for the estimation of sediment concentration runoff transportation in a river system is a mathematical relation between sediment concentration and runoff in a stream or river. The data are applied for functional relation with time lags such as daily, weekly, monthly, seasonally, annually, etc., for study under consideration (Ferguson, 1986). An SRC is constructed mathematically by logarithmic transformation of data and to determine the best fit line using a linear least square regression.

$$C = aQ^b \qquad (12.32)$$

Logarithmic transformation on log-log paper represents a straight line

$$log C = log a + b log (Q) \qquad (12.33)$$

C = sediment runoff (concentration or load), Q = runoff or discharge and a & b are regression constants.

Using the SRC model (equation (12.33)), the sediment rating equations computed for the Pranhita sub-basin for six hydrological gauging stations taken up for the study are shown below:

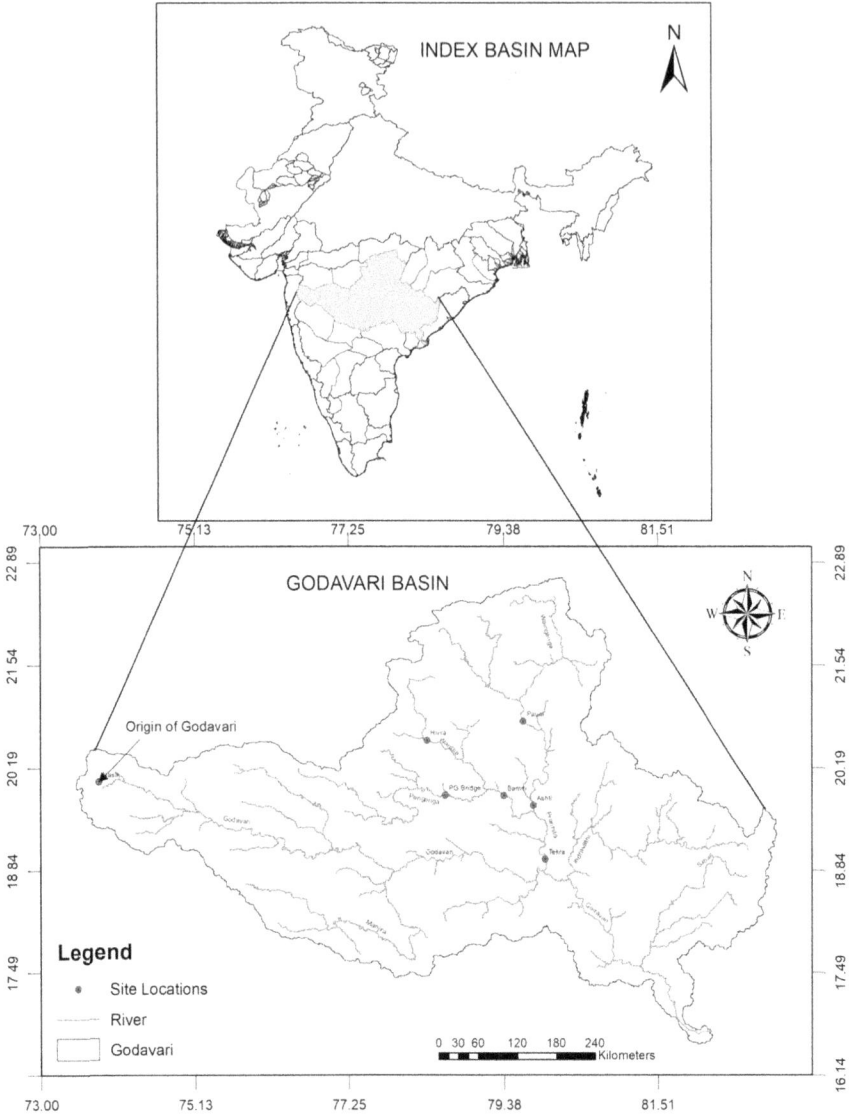

FIGURE 12.4   Major tributary system of Godavari river basin and study locations.

    i. SRC for the Tekra site on main Pranhita River

$$C = 4.940E - 04Q^{0.770} \tag{12.34}$$

    ii. SRC for the Penganga Bridge site on the Penganga River

$$C = 1.671E - 03\ Q^{0.921} \tag{12.35}$$

iii. SRC for the Bamni site on the Wardha River

$$C = 1.154E - 03\ Q^{0.773} \tag{12.36}$$

iv. SRC for the Hivra site on the Wardha River

$$C = 2.000E - 05\ Q^{1.266} \tag{12.37}$$

v. SRC for the Ashti site on the Wainganga River

$$C = 1.280E - 02\ Q^{0.375} \tag{12.38}$$

vi. SRC for the Pauni site on the Wainganga River

$$C = 2.480E - 03\ Q^{0.672} \tag{12.39}$$

where C = sediment concentration (g/lit) and Q = runoff or discharge in river (cumec)

The statistical analysis of ANN models along with SRC models are summarized in Table 12.2 for the six gauging stations, and the best performing models of the respective stations were compared with their SRC technique to show the superiority and accuracy in the modeling process. As presented in Table 12.2, the RMSE values of SRCs are highly deviated in their magnitude, performing very poorly when compared with respective ANN models of all six gauging stations in the study. In the statistic of the correlation coefficient, most of the ANN models performed (more than 0.85, i.e., 85%), and the values are very high in some cases, i.e., more than 95%. Further, all the ANN models have shown good generalization capability in the training phase, testing phase and validation phase when compared with the SRC conventional method. The developmental increment or the improvement in "R" statistic in the validation phase reveals the capability of the model for good generalization. However, at the gauging station Hivra and Pauni, the ANN methodology showed less performance than the other gauging stations as they are comparatively farther locations from the outlets of the stream, having uneven sediment transportation.

The performance of the rating curve model is very normal in "R" statistic and significantly less efficient in the three criteria presented because the calculated sediment values from the SRC model follow a general trend; thereby, higher magnitude R values may result. However, there is significant numerical deviation in observed and estimated sediment concentration. Hence, the values of RMSE and DC are inefficient for the SRC study. Figures 12.5 and 12.6 show the Tekra gauging station of the Pranitha river, where the combined flows of all the gauging stations met, and give a good understanding in the graphical representations of observed and estimated sediment concentration and their scatter plots in (a) training, (b) testing and (c) validation. In Figure 12.5, a small amount of mismatch may be seen between the targeted and calculated sediment concentration series during training,

TABLE 12.2

Performance of Runoff-Sediment ANN Models and SRC Method for Six Gauging Stations of the Study

ANN Model	Network Architecture	Training			Testing		Validation	
		RMSE	R	DC	R	DC	R	DC
Pranhita (Tekra)	[7-6-1]	0.060	0.986	0.973	0.953	0.860	0.943	0.806
Penganga Bridge	[7-6-1]	0.139	0.890	0.790	0.788	0.617	0.924	0.853
Wardha (Bamni)	[3-3-1]	0.237	0.942	0.887	0.887	0.785	0.766	0.572
Wardha (Hivra)	[9-7-1]	0.146	0.844	0.698	0.909	0.785	0.637	0.610
Wainganga (Ashti)	[7-6-1]	0.074	0.945	0.888	0.945	0.881	0.944	0.888
Wainganga (Pauni)	[5-4-1]	0.102	0.938	0.879	0.741	0.252	0.865	0.733
Pranhita (Tekra)	[7-6-5-1]	0.076	0.978	0.957	0.959	0.904	0.954	0.891
Penganga Bridge	[7-6-5-1]	0.150	0.870	0.757	0.830	0.660	0.923	0.851
Wardha (Bamni)	[1-2-5-1]	0.365	0.858	0.732	0.881	0.770	0.743	0.548
Wardha (Hivra)	[5-4-5-1]	0.110	0.940	0.836	0.875	0.567	0.648	0.551
Wainganga (Ashti)	[5-4-6-1]	0.068	0.995	0.906	0.942	0.874	0.930	0.861
Wainganga (Pauni)	[7-6-5-1]	0.122	0.916	0.826	0.723	0.358	0.861	0.733
SRC-Tekra site	-	0.472	0.923	0.680	0.870	0.870	0.867	0.668
SRC-Penganga bridge	-	0.271	0.734	0.204	0.637	0.178	0.785	0.269
SRC-Bamni site	-	0.582	0.710	0.360	0.859	0.473	0.756	0.434
SRC- Hivra		0.263	0.405	0.058	0.914	0.713	0.761	0.404
SRC-Ashti	-	0.154	0820	0.559	0.884	0.715	0.897	0.530
SRC-Pauni	-	0.205	0.865	0.596	0.802	0.644	0.820	0.567

testing and validation. In the same lines of presentation, the scatter plots demonstrate very good correlation values and very negligible or insignificant deviation from the ideal line; otherwise, the deviation from the ideal line may cause systematic errors. It can be seen from Figure 12.5 that the scatter plots in the training phase, testing phase and validation phase show very high values of correlation, and the line follows the ideal line. The comparative plots of temporal variation among the observed time series, ANN estimated series and SRC estimated series are graphically shown in Figure 12.6. ANN methodology followed the trend of the observed time series and yielded very close values, whereas SRC couldn't follow the trend of the real-time series and saw a high mismatch, which is synonymous with results observed in literature by Lohani et al. (2007), Kisi (2007a, b), Rai Mathur (2008) and Sarkar et al. (2004 and 2010).

c. **Performance evaluation criteria**

The performance evaluation carried out in the current study used the statistics parameters: root mean square error (RMSE), correlation coefficient (R) and coefficient of efficiency (CE) or coefficient of determination (DC or R^2).

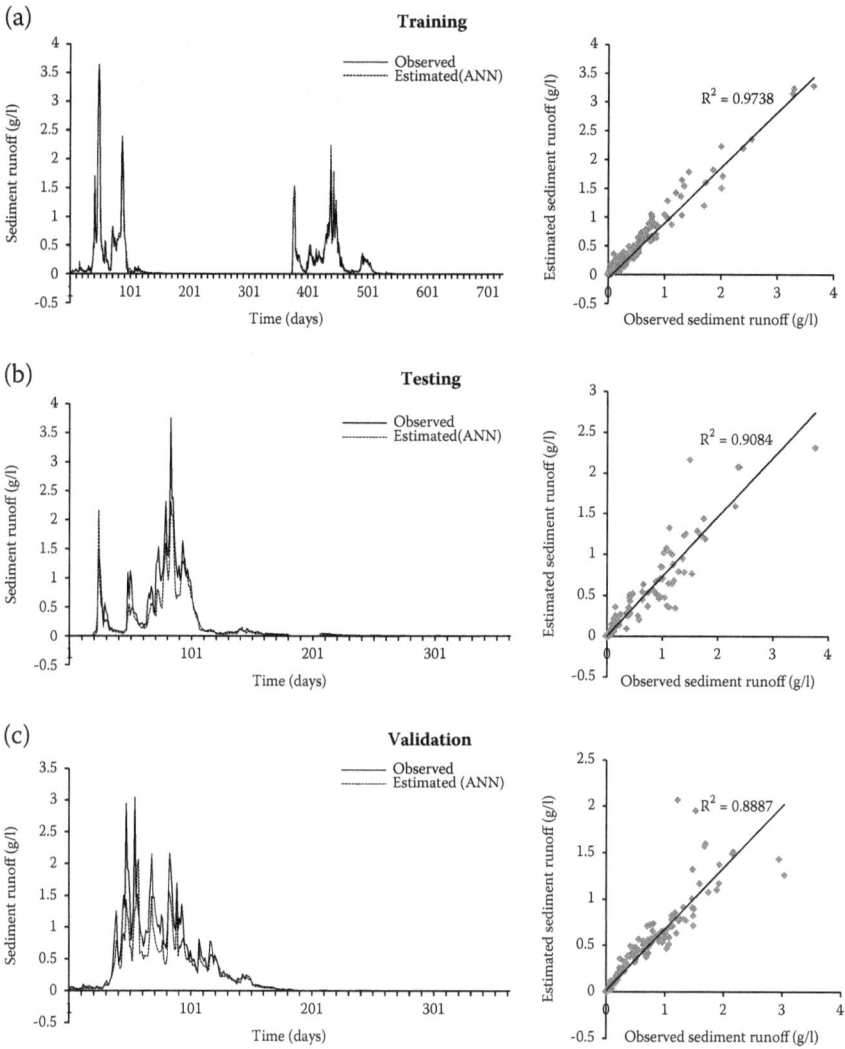

FIGURE 12.5 Targeted and calculated sediment concentration and their scatter plots for the Tekra gauging station of the main Pranhita River.

RMSE evaluates the residual error (Yu et al., 1994) using the following relation:

$$\text{RMSE} = \sqrt{\frac{\text{residual variance}}{n}} = \left(\sum_{j=1}^{n} (Y_j - \hat{Y}_j)^2 / n \right)^{1/2} \qquad (12.40)$$

where Y and $\hat{Y}$ are the observed and estimated values, and "n" is the number of observations.

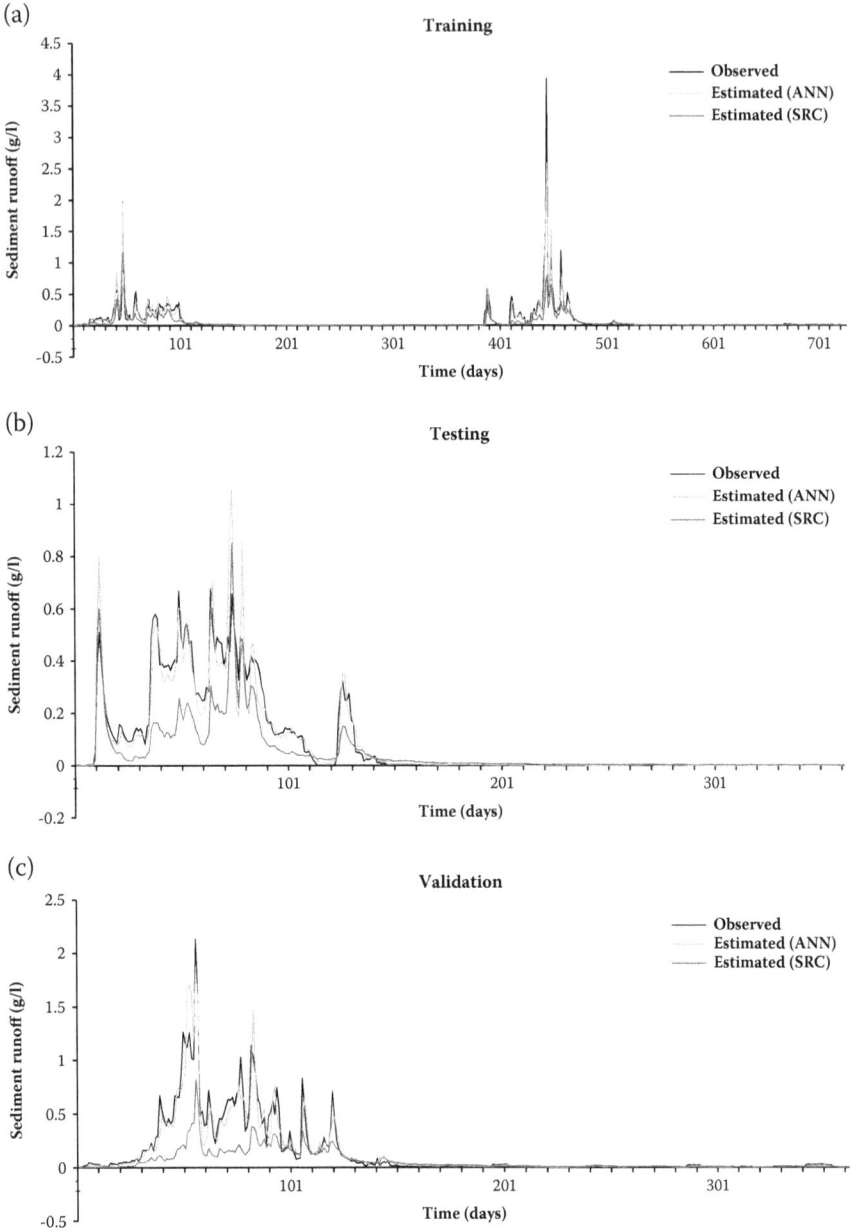

FIGURE 12.6 Observed and estimated (ANN) sediment concentration comparison with SRC sediment concentration at the Ashti gauging station.

Correlation coefficient (R) is expressed as

$$R = \frac{\sum_{j=1}^{n}\left\{\left(Y_j - \bar{Y}\right)\left(\widehat{Y}_j - \bar{\widehat{Y}}\right)\right\}}{\left\{\sum_{j=1}^{n}\left(Y_j - \bar{Y}\right)^2 \sum_{j=1}^{n}\left(\widehat{Y}_j - \bar{\widehat{Y}}\right)^2\right\}^{1/2}} \times 100 \tag{12.41}$$

where $\bar{Y}$ and $\bar{\widehat{Y}}$ are the mean of observed and estimated values.

Determination coefficient (DC)
It is a measure of comparative relative performance between initial variance and standardization of residual variance (Nash & Sutcliffe, 1970). The determination coefficient is synonymously referred to as coefficient of efficiency and represents the fraction of variance that is explained by regression. The value closer to unity indicates better statistical performance of the model under evaluation w.r.t *DC* (Haan, 1977).

$$DC = \left\{1 - \frac{\text{residual variance}}{\text{initial variance}}\right\} \times 100$$

$$= \left\{1 - \frac{\sum_{j=1}^{n}\left(Y_j - \widehat{Y}_j\right)^2}{\sum_{j=1}^{n}\left(Y_j - \bar{Y}\right)^2}\right\} \times 100 \tag{12.42}$$

d. Hysteresis analysis in the runoff-sediment process

The sediment concentration for a specific runoff magnitude on the rising limb of the sediment runoff hydrograph will be larger or smaller than the magnitude of the sediment concentration in the falling limb of the runoff hydrograph, depending upon the severity of the catchment erosion, the amount of sediment that is being transported, the measurement location of the sediment from the source of the sediment, etc. Conventional methodologies such as linear or non-linear regression are unable to comprehend the variations in the relevant processes. The neural network solution capability to capture this natural relationship has been tested and reported in the present study, but the resultant simplifications contained significant under-estimations of important sediment-related events. The study presents some recent findings based on the parameterization of a self-organizing feature map that is used to partition the observed relationship that exists between transported sediment and pertinent fluvial conditions. Hydrological topologies and trajectories are thereafter used to develop a dedicated model that is able to switch between different structural processes and mechanisms and thus offer a suitable construct for incorporation of trigger events. Figures 12.7, 12.8 and 12.9 show real-time series data of sediment concentration combined for the training, testing, and validation phases. The figures

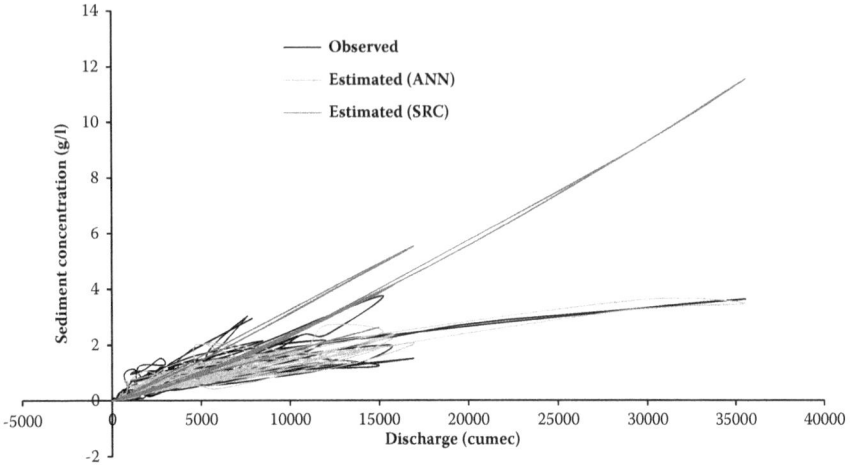

FIGURE 12.7   Observed hysteresis and estimated hysteresis at Tekra (main Pranhita river).

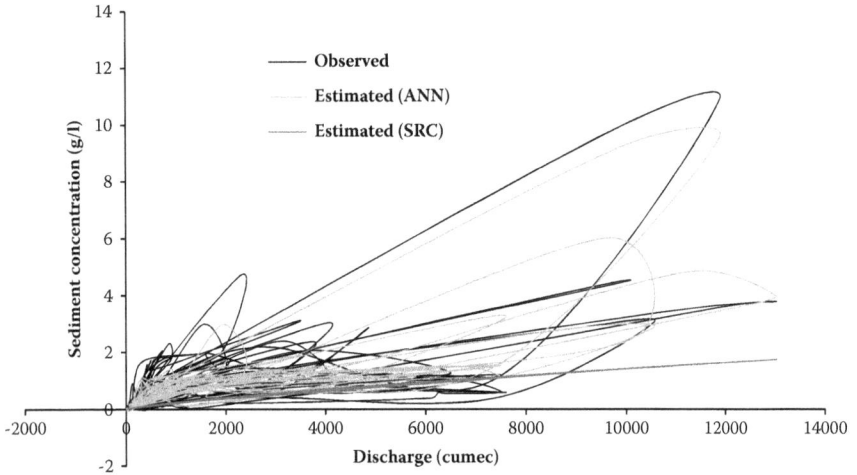

FIGURE 12.8   Observed hysteresis and estimated hysteresis at Bamni (Wardha river).

show that the ANN methodology can describe the events of hydrologic and hydraulic processes like hysteresis with the help of data obtained by ANNs, i.e., simulated and estimated sediment concentration using ANNs and the data estimated
by SRC methodology for the three phases compared in the current study. The
hysteresis estimation by ANNs coincides graphically as well as magnitude-wise
with the hysteresis of observed real-time series data, whereas the SRC is unable to
simulate the hysteresis effect with the observed sediment transport process.

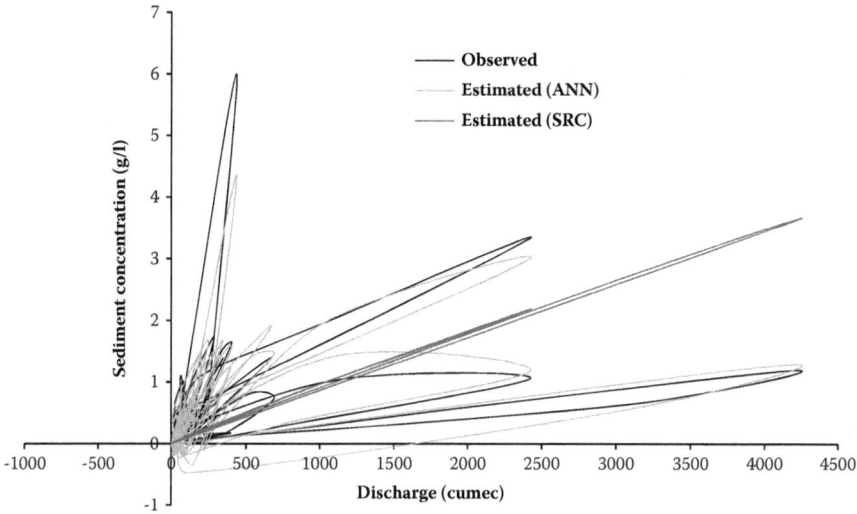

FIGURE 12.9   Observed hysteresis and estimated hysteresis at Hivra (Wardha river).

ii. River Stage Modeling

The study pertaining to the above subject dealt with finding a model/solution to the discharge (q) output in varying river stage (h) situations at a gauging station, Rajahmundry (Dhawalaishwaram Barrage site) in Andhra Pradesh State, India, located on the main Godavari river, downstream and near the outlet of the basin or catchment with the available data on stage-discharge for more than 25 years. Here, the idea of the authors is to analyze the high stages in the study area and to forecast the flood situations in the location, as the location is in frequent flood-prone zone of the river basin. The ANN methodology has been governed to arrive at required solutions in the same guidelines as above in runoff-sediment modeling. In the current study, an attempt has been made for a comparative study with Linear Multiple Regression (LMR) methodology. LMR methodology has given good ideology of the modeling; however, superiority of ANN methodology (Tayfur, 2002) has been shown by acquiring a high level of accuracy. The designed models were trained, tested, validated and finally compared with LMR models to conclude the ANN capability in modeling the river stage— discharge of the Bisht et al. (2010) study with a valid prediction research available on stage by Thirumalaiah and Deo (1998). The performance evaluation was carried out with the statistics RMSE, R, R^2 or DC and mean absolute deviation (MAD), a statistical measure or evaluation of absolute average deviation by the estimated values from real-time series and unit and normalized criterion (Figure 12.10) (Table 12.3).

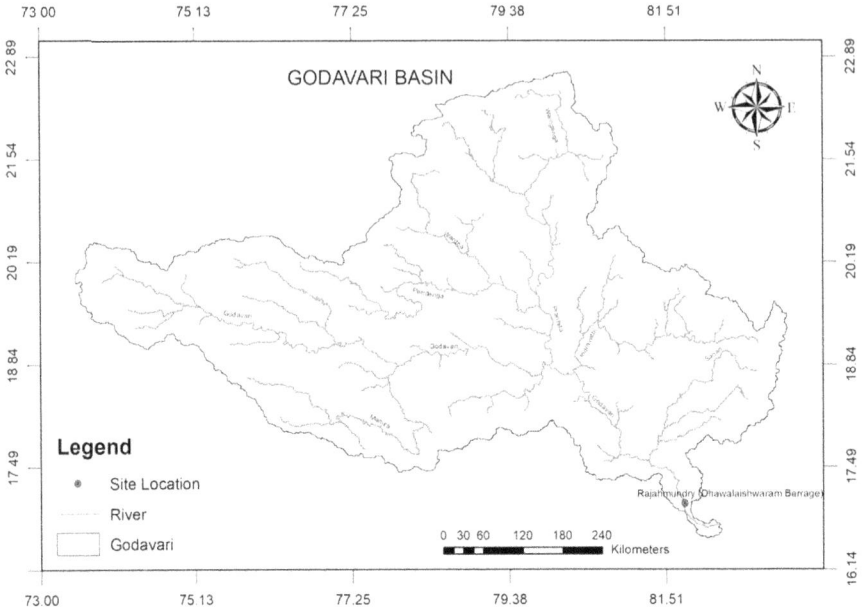

FIGURE 12.10 Hydrological study location.

$$MAD = \frac{\sum_{j=1}^{n} \left| Y_j - \bar{Y}_j \right|}{n} \tag{12.43}$$

LMR Model Design

In LMR design, the River Discharge at time t, (q_t), can be regressed against the River Stage and River Discharge in the past. The LMR models can be represented as

$$q_t = \lambda_0 + \lambda_1 h_t + \lambda_2 h_{t-1} + \lambda_3 h_{t-2} + \lambda_4 h_{t-3} + \lambda_5 q_{t-1} + \lambda_6 q_{t-2} + \lambda_7 q_{t-3} \tag{12.44}$$

where λ_is represents the regression coefficients to be determined; h_is represents the River Stage; q_is represents the Discharge; and t = index representing time.

Designed regression models

Model – LMR-1

$$q_t = 5468.10 + 632.871 h_t + 0.847 q_{t-1} \tag{12.45}$$

Model – LMR-2

$$q_t = 6448.59 + 2988.57 h_t - 2426.495 h_{t-1} + 0.849 q_{t-1} \tag{12.46}$$

TABLE 12.3
Ann Models Design for River Stage and Discharge

Model Number	Hidden Layers Applied	Output	Inputs	Model Number	Hidden Layers Applied	Output	Inputs
ANN-1	One	q_t	h_t	ANN-11	Two	q_t	h_t
ANN-2	One	q_t	h_t, h_{t-1}	ANN-12	Two	q_t	h_t, h_{t-1}
ANN-3	One	q_t	h_t, q_{t-1}	ANN-13	Two	q_t	h_t, q_{t-1}
ANN-4	One	q_t	h_t, h_{t-1}, q_{t-1}	ANN-14	Two	q_t	h_t, h_{t-1}, q_{t-1}
ANN-5	One	q_t	h_t, h_{t-1}, h_{t-2}	ANN-15	Two	q_t	h_t, h_{t-1}, h_{t-2}
ANN-6	One	q_t	h_t, h_{t-1}, h_{t-2}, q_{t-1}	ANN-16	Two	q_t	h_t, h_{t-1}, h_{t-2}, q_{t-1}
ANN-7	One	q_t	h_t, h_{t-1}, h_{t-2}, q_{t-1}, q_{t-2}	ANN-17	Two	q_t	h_t, h_{t-1}, h_{t-2}, q_{t-1}, q_{t-2}
ANN-8	One	q_t	h_t, h_{t-1}, h_{t-2}, h_{t-3}	ANN-18	Two	q_t	h_t, h_{t-1}, h_{t-2}, h_{t-3}
ANN-9	One	q_t	h_t, h_{t-1}, h_{t-2}, h_{t-3}, q_{t-1}, q_{t-2}	ANN-19	Two	q_t	h_t, h_{t-1}, h_{t-2}, h_{t-3}, q_{t-1}, q_{t-2}
ANN-10	One	q_t	h_t, h_{t-1}, h_{t-2}, h_{t-3}, q_{t-1}, q_{t-2}, q_{t-3}	ANN-20	Two	q_t	h_t, h_{t-1}, h_{t-2}, h_{t-3}, q_{t-1}, q_{t-2}, q_{t-3}

Where, q=discharge(cumec), h=stage(meter)

Model – LMR-3

$$q_t = 7152.94 + 3645.588h_t - 970.85h_{t-1} - 2159.02h_{t-2} + 0.849q_{t-1} \quad (12.47)$$

Model – LMR-4

$$q_t = 5830.83 + 4016.82h_t - 835.42h_{t-1} - 2732.44h_{t-2} + 0.716q_{t-1} + 0.157q_{t-2} \quad (12.48)$$

Model – LMR-5

$$q_t = 6033.87 + 4090.09h_t - 604.17h_{t-1} - 2024.75h_{t-2} - 1123.0h_{t-3} + 0.714q_{t-1}$$
$$+ 0.158q_{t-2} \quad (12.49)$$

The best representative models for the hydrological gauging station under study and their comparative performance has been discussed in tabulated form in the following table, and the study concluded to have the superiority of the ANN

TABLE 12.4
Comparison of Best ANN River Stage-Discharge Models with LMR Models

S. No.	Model	MAD (m³/s)	R	R²	RMSE (m³/s)
1	ANN – 9	15535.43	0.938	0.880	28674.17
2	ANN – 13	15097.49	0.947	0.897	27064.17
3	ANN – 14	15168.80	0.944	0.891	27819.76
4	ANN –16	15579.27	0.942	0.890	28298.94
5	ANN – 17	14795.55	0.938	0.880	28821.87
6	LMR – 1	26694.78	0.853	0.728	69422.48
7	LMR – 2	26765.64	0.853	0.729	69311.67
8	LMR – 3	26801.61	0.854	0.730	69245.95
9	LMR – 4	27341.82	0.858	0.736	68389.09
10	LMR – 5	27407.36	0.858	0.736	68385.07

methodology. The study ultimately concluded to show that the performance of the ANN is remarkable, and it is the simple way to predict and forecast the high stages in the vicinity of the flood plains and the methodology sufficient enough to model the system (Table 12.4).

iii. ANN Models for Spring Discharge

The study is an application of ANNs in predicting the weekly spring discharge based on the weekly spring discharge data collected from a catchment having an area of 871 ha (8.71 km²) that discharges outflows into the Henval river in the *Tehri Garhwal* district of Uttarakhand state lies between $78°$ $22'$ $28''$ to $78°$ $24'$ $57''$E longitude and $30°$ $17'$ $19''$ to $30°$ $18'$ $52''$N latitude in India at an altitude variation from 960 to 2,000 meters in elevation above MSL (mean sea level). The spring location under study in particular lies between $78°$ $24'$ $34''$E and $30°$ $18'$ $47''$N at an altitude elevation of 1,844 m mean sea level in dense forest area. The catchment area was tracked with GPS, and the study spring was noted with the help of GPS by locating the same area with GIS environment. Finally, to get the height of the natural spring above MSL, the spring layer was overlaid with DEM. The research has been carried out (Raju et al., 2011) with a prime objective to estimate the spring discharge by simulating and modeling the weekly time series data pertaining to spring discharge (Q), rainfall (R), evaporation (E) and temperature (T) procured for analysis for seven years from 1999 to 2005. The region under study is fully blessed by number of natural springs that serve the purpose of drinking water needs of the people at most in the vicinity of the springs; hence, the research feels a typical estimation of spring water resource will play crucial role in executing the water conservation activities of the region as a whole.

Five models were designed in the prediction of the weekly spring discharge using average weekly data of rainfall, evaporation and temperature with a specific time interval or timestep. These models were trained with quick propagation, batch back propagation and Levenberg Marquardt algorithms using weekly data from 1999 to 2005. A multi-layered feed-forward neural network was employed among the nodes with one and two hidden layers by considering the output as spring discharge at the timestep Q_t in addition to the current values of discharge and discharge at previous timesteps. Therefore, Q_t was mapped with the other variable such as Q_{t-1}, Q_{t-2}, R_{t-1}, R_{t-2}, E_{t-1}, E_{t-2} and T_{t-1}, T_{t-2} to train ANN in the study. As per the guideline of ASCE (2000a) and ASCE (2000b) in one hidden layer architecture the data of 239 weeks used for training, 55 weeks for testing and 55 weeks for validation. Similarly, in the two hidden layer architecture, the data of 234 weeks used for training, 54 weeks for testing and 54 weeks for validation. Normalization was executed in the range of 0 to 1 to avoid the saturation effect by using the sigmoidal transfer activation function in the study (Table 12.5).

TABLE 12.5
ANN Design for Spring Discharge

Model Number	Model Inputs	Hidden Layers Applied	Algorithm Applied	Output
ANN-1	R_t, E_t, T_t	One	Batch backpropagation ($\eta = 0.8$,	Q_t
ANN-2	Q_{t-1}, R_t, R_{t-1}, E_t, E_{t-1}, T_t, T_{t-1}	One	$\alpha = 0.6$ to 0.9) Quick propagation	Q_t
ANN-3	Q_{t-1}, Q_{t-2}, R_t, R_{t-1}, R_{t-2}, E_t, E_{t-1}, E_{t-2}, T_t, T_{t-1}, T_{t-2}	One	(QP coeff = 1.75, $\eta = 0.8$)	Q_t
ANN-4	Q_{t-1}, Q_{t-2}, Q_{t-3}, R_t, R_{t-1}, R_{t-2}, R_{t-3}, E_t, E_{t-1}, E_{t-2}, E_{t-3}, T_t, T_{t-1}, T_{t-2}, T_{t-3}	One		Q_t
ANN-5	Q_{t-1}, Q_{t-2}, Q_{t-3}, Q_{t-4}, R_t, R_{t-1}, R_{t-2}, R_{t-3}, R_{t-4}, E_t, E_{t-1}, E_{t-2}, E_{t-3}, E_{t-4}, T_t, T_{t-1}, T_{t-2}, T_{t-3}, T_{t-4}	One		Q_t
ANN-6	R_t, E_t, T_t	Two		Q_t
Ann-7	Q_{t-1}, R_t, R_{t-1}, E_t, E_{t-1}, T_t, T_{t-1}	Two		Q_t
ANN-8	Q_{t-1}, Q_{t-2}, R_t, R_{t-1}, R_{t-2}, E_t, E_{t-1}, E_{t-2}, T_t, T_{t-1}, T_{t-2}	Two	Levenberg Marquardt (local minima avoidance)	Q_t
ANN-9	Q_{t-1}, Q_{t-2}, Q_{t-3}, R_t, R_{t-1}, R_{t-2}, R_{t-3}, E_t, E_{t-1}, E_{t-2}, E_{t-3}, T_t, T_{t-1}, T_{t-2}, T_{t-3}	Two		Q_t
ANN-10	Q_{t-1}, Q_{t-2}, Q_{t-3}, Q_{t-4}, R_t, R_{t-1}, R_{t-2}, R_{t-3}, R_{t-4}, E_t, E_{t-1}, E_{t-2}, E_{t-3}, E_{t-4}, T_t, T_{t-1}, T_{t-2}, T_{t-3}, T_{t-4}	Two		Q_t

TABLE 12.6

Representative Models for the Study

Model Inputs	Architecture	ANN Quick Propagation		LMR Model	
		R	DC	R	DC
Q_{t-1}, R_t, R_{t-1}, E_t, E_{t-1}, T_t, T_{t-1}	[7-7-1]	0.983	0.964	0.970	0.941
Q_{t-1}, Q_{t-2}, R_t, R_{t-1}, R_{t-2}, Et, E_{t-1}, E_{t-2}, T_t, T_{t-1}, T_{t-2}	[11-6-5-1]	0.990	0.960	0.976	0.948

The LMR model was developed with Q_t as a criterion variable, and the rest variables rainfall, evaporation, temperature and previous timestep spring discharge were used as predictor variables.

$$Q_t = \lambda_0 + \lambda_1 Q_{t-1} + \lambda_2 Q_{t-2} + \lambda_3 R_t + \lambda_4 R_t - 1 + \lambda_5 R_t - 2 + \lambda_6 E_t + \lambda_7 E_{t-1}$$
$$+ \lambda_8 E_{t-2} + \lambda_9 T_t + \lambda_{10} T_{t-1} + \lambda_{11} T_{t-2} \tag{12.50}$$

where λ_is represents the regression coefficients to be determined, R_is represents the rainfall, E_is represents the evaporation, Tis represents the temperature, Q_is represents the spring discharge, and "t" is the index representing time. The following are the regression models designed for the study.

Model: LMR-1:

$$Q_t = 373.56 + 0.91 Q_{t-1} + 107.46 R_t + 248.80 R_{t-1} - 541.81 E_t - 3102.35 E_{t-1}$$
$$+ 1163.65 T_t - 558.72 T_{t-1} \tag{12.51}$$

Model: LMR-2:

$$Q_t = 3870.08 + 1.14 Q_{t-1} - 0.24 Q_{t-2} + 88.84 R_t + 205.11 R_{t-1} + 35.56 R_{t-2}$$
$$- 1097.86 E_t - 563.58 E_{t-1} - 2935.29 E_{t-2} + 992.81 T_t - 289.59 T_{t-1}$$
$$- 2.05 T_{t-2} \tag{12.52}$$

Various trials were made, and the best results are shown in Table 12.6. Almost all the models developed resulted in good generalization capability, including LMR models. However, ANN models were shown fine-tuned statistic criterion values over LMR models. Hence, ANN may be easily adoptable methodology as it is a quick and accurate results-oriented methodology. The graphical presentation of the above results gives a simple and lucid idea of the methodology in good understanding as follows (Figure 12.11).

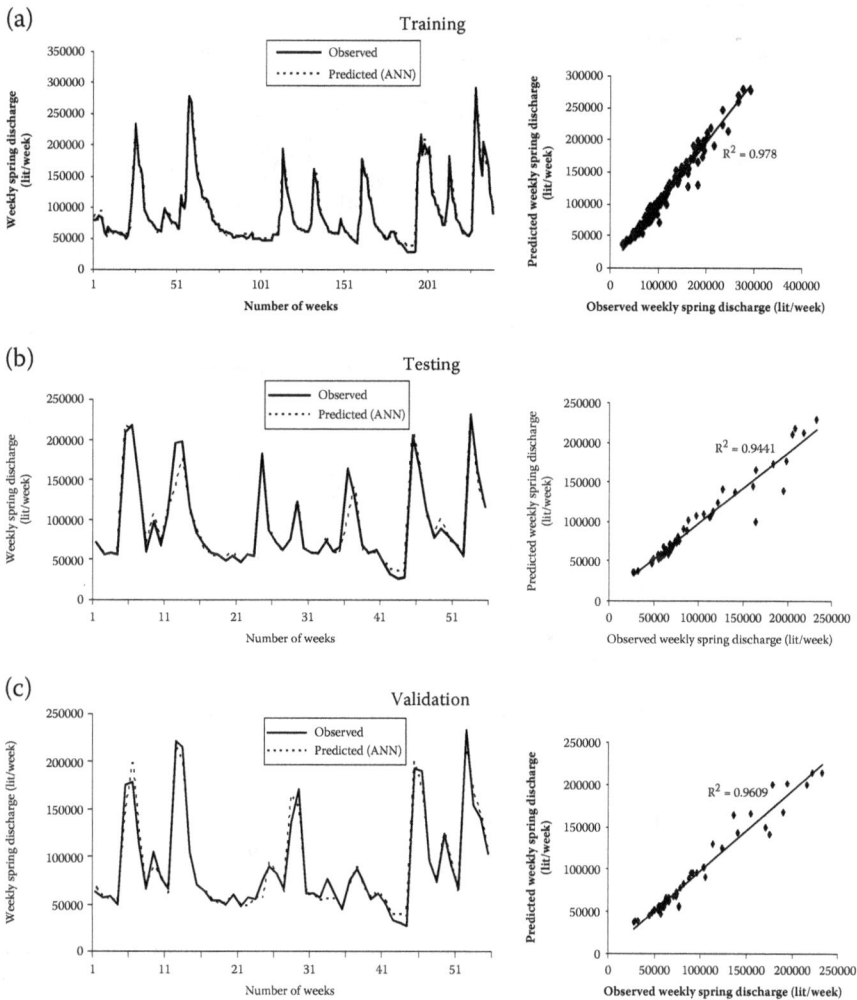

FIGURE 12.11 Weekly spring discharge of real time series and calculated by ANN.

iv. Groundwater—Water table Modeling with Fuzzy Logic and ANN

Groundwater is a major drinking water source in the region under study, and its estimation in respect to elevation fluctuation is a major concern for optimal usage of the resource. Hydro-geological studies many times will forecast the water table depth or elevation, and it does not require a very precise measurement all the time; however, a precise model is a complicated and uneconomical element in the development process when it is required for simulation (Bisht et al., 2009; 2013; Lohani et al., 2006). The study area (5163 km^2) is a part of the Gangetic alluvial plain with flat topography and slopes from northwest to southeast of the Ganga-Ramganga inter-basin, located geographically at the longitudes of 78^0 15' and 79^0

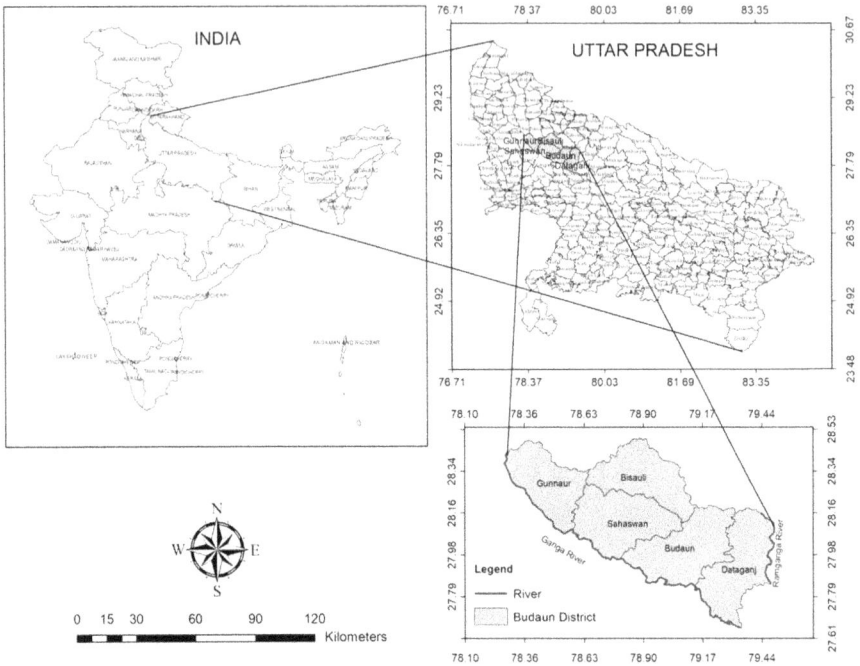

FIGURE 12.12 Location of the groundwater study (Budaun district index map).

30′ E; and latitudes of 27^0 30′ and 28^0 30′ N in the Budaun district of the Uttar Pradesh state of India. MATLAB 7.0 was applied to execute the prediction process by using fuzzy logic and ANN. The performance evaluation was carried out with parameters such as R, DC(R^2), RMSE, MAD and coefficient of efficiency (CE) (Figure 12.12).

Model development was executed by considering groundwater recharge and groundwater discharge in the current year as input, i.e., at timestep t for an output of water table elevation in the current year at timestep t in the design of model 1. Subsequently, in model 2, an additional input of water table elevation of the previous year post monsoon at timestep t-1 was considered along with two other inputs considered in the model 1 as inputs. Further, in model 3 previous year post monsoon recharge was considered as an input, i.e., at timestep t-1 in addition to current year recharge and discharge, and so on as followed in Table 12.7. Likewise five models were designed for fuzzy logic and ten models for ANN (five with one hidden layer and five with two hidden layers). Fuzzy if-then rules imposed among the variables of the rule-based fuzzy systems i.e. "if antecedent proposition then consequent proposition".

Many numbers of trials were made to model the system behavior with ANN and fuzzy logic and found that all designed model have shown very good statistical performance with high correlation and regression coefficients; more than 90% showed good generalization capacity (Bisht et al., 2013). Hence, the study comprehends the effective performance of the soft computing and discovers the

TABLE 12.7

Designed Models of ANN and Fuzzy Logic

S.No.	Model	No. of Hidden Layers	Output	Input Variables
1	ANN – 1	One	W_t	R_t, D_t
2	ANN – 2	One	W_t	R_t, D_t, W_{t-1}
3	ANN – 3	One	W_t	R_t, D_t, R_{t-1}
4	ANN – 4	One	W_t	$R_t, D_t, R_{t-1}, D_{t-1}$
5	ANN – 5	one	W_t	$R_t, D_t, R_{t-1}, D_{t-1}, W_{t-1}$
6	ANN – 6	Two	W_t	R_t, D_t
7	Ann – 7	Two	W_t	R_t, D_t, W_{t-1}
8	ANN – 8	Two	W_t	R_t, D_t, R_{t-1}
9	ANN – 9	Two	W_t	$R_t, D_t, R_{t-1}, D_{t-1}$
10	ANN – 10	Two	W_t	$R_t, D_t, R_{t-1}, D_{t-1}, W_{t-1}$
11	Fuzzy – 1	-	$W_t =$	$f\{R(t), D(t)\}$
12	Fuzzy – 2	-	$W_t =$	$f\{R(t), D(t), W(t-1)\}$
13	Fuzzy – 3	-	$W_t =$	$f\{R(t), D(t), R(t-1)\}$
14	Fuzzy – 4	-	$W_t =$	$f\{R(t), D(t), R(t-1), D(t-1)\}$
15	Fuzzy – 5	-	$W_t =$	$f\{R(t), D(t), R(t-1), D(t-1), W(t-1)\}$

suitability of soft computing for quick and emergent prediction/forecasting situations, unlike conventional techniques of the hydrologic process time series as the conventional method needs huge amount of data for the analysis. For a good comparative study in the following Table 12.8, the splendid performance of the four models has been presented. It is also observed from the table that both in the case of ANN and fuzzy logic modeling model 2 performed highly superior to the others. In model 2 the inputs are R_t, D_t and W_{t-1}. Further, the study recommends that fuzzy logic and ANN methodologies are the best hydrologic simulation and modeling techniques.

v. Groundwater—Water table modeling with Fuzzy Logic and ANFIS

The authors' insight into the present trial is a light blend of hybrid modeling by using fuzzy logic and Adaptive Neuro Fuzzy Inference System (ANFIS) in the present investigation in the context of their applied time series neural network methodologies. Keeping this in view, fuzzy logic was used to observe the vagueness or fuzziness to handle the uncertainty of the system, and ANFIS was applied to improve generalized capacity in the modeling process (Bisht et al., 2009). Usual groundwater models are mathematically obtained by suitable differential equations to describe the behavior due to complexity and nonlinear features of the system. The study area (5163 km^2) is a part of the Gangetic alluvial plain with flat topography and slopes from northwest to southeast of the Ganga-Ramganga inter-basin

TABLE 12.8
Best Performed Models of the Designed Methodology

S.No.	Model	No. of Hidden layer	R	R^2	MAD	RMSE
1	ANN-1	One	0.974351	0.949	2.000000	0.622178
2	ANN-2	One	0.977273	0.955	0.292719	0.531218
3	Fuzzy-1	-	0.985183	0.970	0.360000	0.437822
4	Fuzzy-2	-	0.993646	0.987	0.260547	0.317821

located geographically at the longitudes of 78^0 15′ and 79^0 30′ E; and latitudes of 27^0 30′ and 28^0 30′ N in the Budaun district of the Uttarpradesh state of India. MATLAB 7.0 was applied to execute the prediction process by using the fuzzy logic toolbox in designing the models. By using observed data, membership functions were constructed, and FIS was prepared using fuzzy logic toolbox in MATLAB. The rules framed with observed data were then added to fuzzy logic rule editor. The FIS was then evaluated to determine the output.

a. Fuzzification methodology

The data to be studied—groundwater recharge, groundwater discharge, previous groundwater table elevation above MSL and present groundwater table elevation above MSL—were fuzzified into fuzzy subsets as the data were qualitative rather to linguistic. The linguistic level of the participating modeling variables based on the subjective perception was defined as mentioned in Table 12.7. Defuzzification has been employed to get a crisp output from a fuzzy set result obtained from an implication.

Centroid method of defuzzification was used in the study and is given by an algebraic expression as below: (Table 12.9)

$$Z^* = \frac{\int \mu_c(z) \cdot z dz}{\int \mu_c(z) dz} \tag{12.53}$$

b. ANFIS methodology

Sugeno-type FIS, a combination of a fuzzy inference system and an adaptive neural network was used for prediction of water table elevation fluctuation. ANFIS primarily consists of five components: inputs data base, output data base, fuzzy system generator, an FIS and adaptive neural network.

Model 1: The model was developed by considering the groundwater recharge and groundwater discharge as prime inputs to model output as water table elevation, i.e., fuzzy logic rule base model (FLM1) and ANFIS 1 were used in the prediction process.

TABLE 12.9
Fuzzy Subset Linguistic Design

S.No.	Modeling Variable	Linguistic Level	No. of Subsets
1.	Groundwater Recharge	Very low (VL), low (L), medium (M), high (H) and very high (VH)	5
2.	Groundwater Discharge	Very low (VL), low (L), medium (M), high (H), very high (VH) and extremely high (EH)	6
3.	Water Table Elevation (Input)	Extremely low (EL), very low (VL), low (L), medium (M), high (H), very high (VH) and extremely high (EH)	7
4.	Water Table Elevation (Output)	Extremely low (EL), very low (VL), low (L), medium (M), high (H), very high (VH) and extremely high (EH)	7

Model 2: This model was developed by the same concept used in Model 1 with an additional input variable as water table elevation at timestep *t–1* means post monsoon's water table elevation i.e. Here t represents the current year data and *t–1* represents the previous year's data. FLM 2 and ANFIS 2 were used in the prediction process.

c. **Results**

Results have shown a very close agreement between observed and estimated values. The fuzzy logic approach and ANFIS approach both yielded very good predictions to a maximum level; however, ANFIS was extraordinarily resulted to report and present system scenario having accommodating to greatest regression coefficient value 1, i.e., 100% during ANFIS model 2 with five subsets of membership functions. Hence, the research could predict the system very precisely when modeled with ANFIS. Comparatively, when working with fuzzy logic, ANN and ANFIS for the simulation and modeling of groundwater elevation, researchers have experienced highly efficient mathematical and statistical performance. Further, fine performance has been observed when applying the hybrid model.

12.5 CONCLUSIONS

This chapter contains comprehensive research carried out by the authors out of their interest to study ANN application using hydrological time series in soil and water resources engineering after having gone through the literature available from ASCE (2000a) and ASCE (2000b). In the process, the authors collected possible data for application from relevant literature and various sources like the Central Water

Commission (CWC), New Delhi; National Institute of Hydrology, Roorkee; Department of Soil and Water Conservation Engineering; and Department of Irrigation and Drainage Engineering at The G.B. Pant University of Agriculture & Technology, Pantnagar, in India. The chapter will be helpful for a beginner as a basic key note and study material in the design and development of advanced modeling techniques like ANNs, fuzzy logic, ANFIS, etc., as how to select the input and output parameters along with dependent and independent variable parameters while working with the simulation process. The study also reveals that the hysteresis process can only be described with the help of soft computing as the conventional method is unable to explain the process. At this juncture, we mention and acknowledge that the different tools used for the current research, i.e., trial version software, etc., available online are neural power for sediment transport, alyuda neuroIntelligence for river stage and spring discharge and MATLAB 7.0 for groundwater table elevation.

REFERENCES

Agarwal, A., Mishra, S. K., Ram, S., & Singh, J. K. (2006). Simulation of runoff and sediment yield using artificial neural networks. *Biosystems Engineering*, *94*(4), 597–613.

Anmala, J., Zhang, B., & Govindaraju, R. S. (2000). Comparison of ANNs and empirical approaches for predicting watershed runoff. *Journal of Water Resources Planning and Management, ASCE, 126*(3), 156–166.

ASCE Task Committee. on Application of Artificial Neural Networks in hydrology. (2000a). Artificial neural networks in hydrology. I: Preliminary concepts. *Journal of Hydrologic Engineering, ASCE, 5*(2), 115–123.

ASCE, Task committee on Application of Artificial Neural Networks in Hydrology. (2000b). Artificial neural networks in hydrology. II: Hydrologic applications. *Journal of Hydrologic Engineering, ASCE, 5*(2), 124–137.

Bisht, D. C. S., Raju, M. M., & Joshi, M. C. (2009). Simulation of water table elevation fluctuation using Fuzzy-Logic and ANFIS. *Computer Modelling and New Technologies, 13*(2), 16–23.

Bisht, D. C. S., Raju, M. M., & Joshi, M. C. (2010). ANN based river stage – discharge modelling for Godavari river, India. *Computer Modelling and New Technologies, 14*(3), 48–62.

Bisht, D., Jain, S., & Raju, M. M. (2013). Prediction of water table elevation fluctuation through Fuzzy Logic & Artificial Neural Networks. *International Journal of Advanced Science and Technology, 51*, 107–120.

Cigizoglu, H. K. (2002). Suspended sediment estimation and forecasting using artificial neural networks. *Turkish Journal of Engineering and Environmental Sciences, 26*(2002), 15–25.

Day, S. P., & Davenport, M. R. (1993). Continuous time temporal back-propagation with adaptable time delays. *IEEE Transactions on Neural Networks, 4*(2), 348–354.

Ferguson, R. I. (1986). River loads underestimated by rating curves. *Water Resources Research, 22*(1), 74–76.

Govindaraju, R. S., & Kavvas, M. L. (1991). Modeling the erosion process over steep slopes: Approximate analytical solutions. *Journal of Hydrology, 127*(1–4), 279–305.

Haan, C. T. (1977). *Statistical methods in hydrology.* The Iowa State University Press.

Hario, H., & Jokinen, P. (1991). Increasing the learning speed of back-propagation algorithm by linearization. In T. Kohonen, K. Makisara, O. Simula, & J. Kangas Eds. *Artificial neural networks* (pp. 629–634). Elsevier.

Haykin, S. (1994). *Neural networks – A comprehensive foundation*. Mcmillan.

Jain, S. K. (2008). Development of integrated discharge and sediment rating relation using a compound neural network. *Journal of Hydrologic Engineering, ASCE, 13*(3), 124–131.

Karnin, E. D. (1990). A simple procedure for pruning back propagation trained neural netoworks. *IEEE Transaction on Neural Networks, 1*, 239–242.

Karunanithi, N., Grenney, W. J., Whitley, D., & Bovee, K. (1994). Neural networks for river flow prediction. *Journal of Computing in Civil Engineering, ASCE, 8*(2), 201–220.

Kisi, O. (2007a). Streamflow forecasting using different artificial neural network algorithms. *Journal of Hydrologic Engineering, ASCE, 12*(5), 532–539.

Kisi, O. (2007b). Development of streamflow-suspended sediment rating curve using a range dependent neural network. *International Journal of Science & Technology, 2*(1), 49–61.

Kohonen, T. (1998). *Self organization and associative memory* (2nd Ed.). Springer Verlag.

Kothari, R., & Agyepong, K. (1996). On lateral connections in feedforward neural networks. In Proceedings of *IEEE International Conference on Neural Networks* (pp. 13–18). Institute of Electrical and Electronics Engineers.

Lohani, A. K., Goel, N. K., & Bhatia, K. K. S. (2006). Takagi-Sugeno fuzzy inference system for modeling stage-discharge relationship. *Journal of Hydrology, 331*, 146–160.

Lohani, A. K., Goel, N. K., & Bhatia, K. K. S. (2007). Deriving stage-discharge-sediment concentration relationships using fuzzy logic. *Hydrological Sciences – Journal-des Sciences Hudrologiques, 52*(4), 793–807.

McCulloch, W. S., & Pitts, W. (1943) A logic Calculus of the ideas immanent in nervous activity. *Bulletin of Mathematical Biophysics, 5*, 115–133.

Nadi, F. (1991). Topological design of modular neural networks. In T. Kohonen, K. Makisara O. Simula, & J. Kangas (Eds.), *Artificial neural networks* (pp. 213–217). Elsevier.

Naresh, A., Raju, M. M., Naik, M. G., Gupta, H., Kumar, A., Sarkar, A. (2019). Sediment transport modeling and hysteresis study for pranahita sub-basin of godavari river system in india. *Hydro-2019, International conference on Hydraulics, Water Resources & Coastal Engineering*, 18–20, December 2019, Osmania University, Hyderabad, India.

Nash, J. E., & Sutcliffe, J. V. (1970). River flow forecasting through conceptual models. Part 1 – A: Discussion principles. *Journal of Hydrology, 10*, 282–290.

Raghuvanshi, N. S., Singh, R., & Reddy, L. S. (2006). Runoff and sediment yield modeling using artificial neural networks: Upper Siwane River, India. *Journal of Hydrologic Engineering, ASCE, 11*(1), 71–79.

Rai, R. K., & Mathur, B. S. (2008). Event – based sediment yield modeling using artificial neural network. *Water Resources Management, 22*(4), 423–441.

Raju, M. M. (2008). *Runoff – Sediment modeling using artificial neural networks* [M.Tech Thesis, Soil and Water Conservation Engineering & Irrigation and Drainage Engineering, G.B. Pant University of Agriculture and Technology, Pantnagar-263145, Uttarkhand, India].

Raju, M. M., Srivastava, R. K., Bisht, D. C. S., Sharma, H. C., & Kumar, A. (2011). Development of artificial neural-network-based models for the simulation of spring discharge. *Advances in Artificial Intelligence, Hindawi Publishing Corporation, 2011*, 1–11.

Refenes, A. N., & Vithlani, S. (1991). Constructive learning by specialization. In T. Kohonen, K. Makisara, O. Simula, & J. Kangas (Eds.), *Artificial neural networks* (pp. 923–929). Elsevier.

Sarkar, A., Kumar, R., Singh, R. D., Thakur, G., & Jain, S. K. (2004). Sediment-discharge modeling in a river using artificial neural networks. In Proceedings of the *International Conference ICON – HERP India*, Oct 26–28, 2004. IIT Roorkee.

Sarkar, A., Raju, M. M., & Kumar, A. (2010, January). Sediment runoff modeling using artificial neural networks. *Journal of Indian Water Resources Society*, *30*(1), 39–45.

Sato, A. (1991). An analytical study of the momentum term in a back-propagation algorithm. In T. Kohonen, K. Makisara, O. Simula, & J. Kangas (Eds.), *Artificial neural networks* (pp. 617–622). Elsevier.

Tayfur, G. (2002). Applicability of sediment transport capacity models for nonsteady state erosion from steep slopes. *Journal of Hydrologic Engineering, ASCE*, *7*(3), 252–259.

Tayfur, G., & Guldal, V. (2005). Artificial neural networks for estimating daily total suspended sediment in natural streams. *Nordic Hydrology*, *37*(1), 69–79.

Thirumalaiah, K., & Deo, M. C. (1998). River stage forecasting using artificial neural networks. *Journal of Hydrologic Engineering, ASCE*, *3*(1), 26–32.

Wu, T. H., Hall, J. A., & Bonta, J. V. (1993). Evaluation of runoff and erosion models. *Journal of Irrigation and Drainage Engineering ASCE*, *119*(4), 364–382.

Yu, P. S., Liu, C. L., & Lee T. Y. (1994). Application of a transfer function model to a storage runoff process. *Stochastic and Statistical Methods in Hydrology and Environmental Engineering*, *3*, 87–97.

Zurada, M. Z. (1992). Introduction to artificial neural system. West Publishing Company.

Index

For Product Safety Concerns and Information please contact our EU
representative GPSR@taylorandfrancis.com
Taylor & Francis Verlag GmbH, Kaufingerstraße 24, 80331 München, Germany

www.ingramcontent.com/pod-product-compliance
Lightning Source LLC
Chambersburg PA
CBHW060406220326
41598CB00023B/3036

*9 7 8 0 3 6 7 6 0 8 6 9 9 *